Workbook to Accompany Heavy-Duty Truck Systems

Seventh Edition

Sean Bennett

Australia • Brazil • Mexico • Singapore • United Kingdom • United States

Workbook to accompany Heavy Duty Truck Systems, Seventh Edition

Sean Bennett

SVP, GM Skills & Global Product Management: Jonathan Lau

Product Director: Matthew Seeley

Senior Product Manager: Katie McGuire

Product Assistant: Kimberly Klotz

Executive Director of Development: Marah Bellegarde

Learning Design Director: Juliet Steiner

Senior Learning Designer: Mary Clyne

Vice President Strategic Marketing Services: Jennifer Ann Baker

Marketing Director: Sean Chamberland

Associate Marketing Manager: Andrew Ouimet

Senior Director Content Delivery: Wendy Troeger

Manager Content Delivery: Alexis Ferraro

Senior Content Manager: Sharon Chambliss

Senior Digital Delivery Lead: Amanda Ryan

Design Director: Jack Pendleton

Senior Designer: Angela Sheehan

Cover image(s): iStockPhoto.com/EasyBuy4u

Production Service/Composition: SPi Global

© 2020, 2016 Cengage Learning, Inc.

Unless otherwise noted, all content is © Cengage.

ALL RIGHTS RESERVED. No part of this work covered by the copyright herein may be reproduced or distributed in any form or by any means, except as permitted by U.S. copyright law, without the prior written permission of the copyright owner.

> For product information and technology assistance, contact us at
> **Cengage Customer & Sales Support, 1-800-354-9706**
> or **support.cengage.com**.
>
> For permission to use material from this text or product, submit all requests online at **www.cengage.com/permissions**.

Library of Congress Control Number: 2018944693

ISBN: 978-1-3377-8711-6

Cengage
20 Channel Center Street
Boston, MA 02210
USA

Cengage is a leading provider of customized learning solutions with employees residing in nearly 40 different countries and sales in more than 125 countries around the world. Find your local representative at **www.cengage.com**.

Cengage products are represented in Canada by Nelson Education, Ltd.

To learn more about Cengage platforms and services, register or access your online learning solution, or purchase materials for your course, visit **www.cengage.com**.

Notice to the Reader
Publisher does not warrant or guarantee any of the products described herein or perform any independent analysis in connection with any of the product information contained herein. Publisher does not assume, and expressly disclaims, any obligation to obtain and include information other than that provided to it by the manufacturer. The reader is expressly warned to consider and adopt all safety precautions that might be indicated by the activities described herein and to avoid all potential hazards. By following the instructions contained herein, the reader willingly assumes all risks in connection with such instructions. The publisher makes no representations or warranties of any kind, including but not limited to, the warranties of fitness for particular purpose or merchantability, nor are any such representations implied with respect to the material set forth herein, and the publisher takes no responsibility with respect to such material. The publisher shall not be liable for any special, consequential, or exemplary damages resulting, in whole or part, from the readers' use of, or reliance upon, this material.

Printed in the United States of America
Print Number: 02   Print Year: 2019

Contents

ASE Education Foundation Correlation x
Preface xv

Chapter 1   **Introduction to Servicing Heavy-Duty Trucks** 1
     Objectives 1
     Practice Questions 1
     Job Sheet 1.1 3
     Job Sheet 1.2 5
     Job Sheet 1.3 7
     Study Tips 8

Chapter 2   **Shop Safety and Operations** 9
     Objectives 9
     Practice Questions 9
     Job Sheet 2.1 11
     Job Sheet 2.2 13
     Online Tasks 15
     Study Tips 15

Chapter 3   **Tools and Fasteners** 17
     Objectives 17
     Practice Questions 17
     Job Sheet 3.1 19
     Job Sheet 3.2 21
     Job Sheet 3.3 23
     Job Sheet 3.4 29
     Online Tasks 31
     Study Tips 32

Chapter 4   **Maintenance Programs** 33
     Objectives 33
     Practice Questions 33
     Job Sheet 4.1 37
     Job Sheet 4.2 43
     Job Sheet 4.3 53
     Job Sheet 4.4 57
     Job Sheet 4.5 61
     Job Sheet 4.6 67
     Job Sheet 4.7 69

	Job Sheet 4.8 73
	Job Sheet 4.9 79
	Job Sheet 4.10 85
	Job Sheet 4.11 91
	Job Sheet 4.12 97
	Job Sheet 4.13 101
	Job Sheet 4.14 103
	Online Tasks 105
	Study Tips 105

Chapter 5 Fundamentals of Electricity 107
Objectives 107
Practice Questions 107
Job Sheet 5.1 109
Job Sheet 5.2 111
Online Tasks 115
Study Tips 115

Chapter 6 Fundamentals of Electronics and Computers. 117
Objectives 117
Practice Questions 117
Job Sheet 6.1 121
Job Sheet 6.2 125
Online Tasks 127
Study Tips 127

Chapter 7 Batteries. 129
Objectives 129
Practice Questions 129
Job Sheet 7.1 131
Online Tasks 133
Study Tips 133

Chapter 8 Charging Systems. 135
Objectives 135
Practice Questions 135
Job Sheet 8.1 137
Job Sheet 8.2 141
Online Tasks 144
Study Tips 144

Chapter 9 Cranking Systems. 145
Objectives 145
Practice Questions 145

	Job Sheet 9.1	147
	Job Sheet 9.2	151
	Job Sheet 9.3	153
	Online Tasks	156
	Study Tips	156
Chapter 10	**Chassis Electrical Circuits**	**157**
	Objectives	157
	Practice Questions	157
	Job Sheet 10.1	159
	Job Sheet 10.2	161
	Job Sheet 10.3	163
	Online Tasks	166
	Study Tips	166
Chapter 11	**Diagnosis and Repair of Electronic Circuits**	**167**
	Objectives	167
	Practice Questions	167
	Job Sheet 11.1	169
	Job Sheet 11.2	171
	Job Sheet 11.3	173
	Online Tasks	176
	Study Tips	176
Chapter 12	**Multiplexing**	**177**
	Objectives	177
	Practice Questions	177
	Job Sheet 12.1	179
	Job Sheet 12.2	181
	Job Sheet 12.3	183
	Online Tasks	184
	Study Tips	184
Chapter 13	**Hydraulics**	**185**
	Objectives	185
	Practice Questions	185
	Job Sheet 13.1	187
	Online Tasks	188
	Study Tips	188
Chapter 14	**Clutches**	**189**
	Objectives	189
	Practice Questions	189
	Job Sheet 14.1	191
	Job Sheet 14.2	195

    Online Tasks 196
    Study Tips 196

Chapter 15 **Standard Transmissions** **197**
    Objectives 197
    Practice Questions 197
    Job Sheet 15.1 199
    Online Tasks 200
    Study Tips 200

Chapter 16 **Standard Transmission Servicing** **201**
    Objectives 201
    Practice Questions 201
    Job Sheet 16.1 203
    Online Tasks 205
    Study Tips 205

Chapter 17 **Torque Converters** **207**
    Objectives 207
    Practice Questions 207
    Job Sheet 17.1 209
    Online Tasks 212
    Study Tips 212

Chapter 18 **Automatic Transmissions** **213**
    Objectives 213
    Practice Questions 213
    Job Sheet 18.1 215
    Online Tasks 216
    Study Tips 216

Chapter 19 **Automatic Transmission Maintenance** **217**
    Objectives 217
    Practice Questions 217
    Job Sheet 19.1 219
    Job Sheet 19.2 221
    Online Tasks 222
    Study Tips 222

Chapter 20 **Automated Manual and Hybrid Transmissions** **223**
    Objectives 223
    Practice Questions 223
    Job Sheet 20.1 225

Online Tasks 226
Study Tips 226

**Chapter 21 Electronically Controlled Automatic
Transmissions (ECATs) 227**
Objectives 227
Practice Questions 227
Job Sheet 21.1 231
Job Sheet 21.2 233
Job Sheet 21.3 235
Online Tasks 237
Study Tips 237

Chapter 22 Driveshaft Assemblies. 239
Objectives 239
Practice Questions 239
Job Sheet 22.1 241
Online Tasks 243
Study Tips 243

Chapter 23 Heavy-Duty Truck Axles 245
Objectives 245
Practice Questions 245
Job Sheet 23.1 247
Online Tasks 249
Study Tips 249

Chapter 24 Heavy-Duty Truck Axle Service and Repair 251
Objectives 251
Practice Questions 251
Job Sheet 24.1 253
Job Sheet 24.2 255
Online Tasks 258
Study Tips 258

Chapter 25 Steering and Alignment 259
Objectives 259
Practice Questions 259
Job Sheet 25.1 261
Job Sheet 25.2 263
Job Sheet 25.3 265
Job Sheet 25.4 267
Job Sheet 25.5 269
Job Sheet 25.6 271

	Online Tasks	272
	Study Tips	272

Chapter 26 Suspension Systems ... 273
Objectives ... 273
Practice Questions ... 273
Job Sheet 26.1 ... 275
Job Sheet 26.2 ... 281
Job Sheet 26.3 ... 283
Online Tasks ... 286
Study Tips ... 286

Chapter 27 Wheels and Tires ... 287
Objectives ... 287
Practice Questions ... 287
Job Sheet 27.1 ... 291
Job Sheet 27.2 ... 293
Job Sheet 27.3 ... 295
Job Sheet 27.4 ... 297
Online Tasks ... 299
Study Tips ... 299

Chapter 28 Truck Brake Systems ... 301
Objectives ... 301
Practice Questions ... 301
Job Sheet 28.1 ... 303
Job Sheet 28.2 ... 307
Online Tasks ... 310
Study Tips ... 310

Chapter 29 Hydraulic Brakes and Air-over-Hydraulic Brake Systems ... 311
Objectives ... 311
Practice Questions ... 311
Job Sheet 29.1 ... 313
Online Tasks ... 316
Study Tips ... 316

Chapter 30 ABS and EBS ... 317
Objectives ... 317
Practice Questions ... 317
Job Sheet 30.1 ... 319
Online Tasks ... 320
Study Tips ... 320

Chapter 31   Air Brake Servicing 321
    Objectives 321
    Practice Questions 321
    Job Sheet 31.1 323
    Job Sheet 31.2 331
    Job Sheet 31.3 339
    Job Sheet 31.4 341
    Online Tasks 343
    Study Tips 343

Chapter 32   Vehicle Chassis Frame 345
    Objectives 345
    Practice Questions 345
    Job Sheet 32.1 347
    Online Tasks 348
    Study Tips 348

Chapter 33   Heavy-Duty Truck Trailers 349
    Objectives 349
    Practice Questions 349
    Job Sheet 33.1 351
    Job Sheet 33.2 353
    Online Tasks 355
    Study Tips 355

Chapter 34   Fifth Wheels and Coupling Systems 357
    Objectives 357
    Practice Questions 357
    Job Sheet 34.1 359
    Job Sheet 34.2 361
    Job Sheet 34.3 365
    Online Tasks 366
    Study Tips 366

Chapter 35   Heavy-Duty Heating, Ventilation, and Air-Conditioning Systems 367
    Objectives 367
    Practice Questions 367
    Job Sheet 35.1 369
    Job Sheet 35.2 371
    Online Tasks 373
    Study Tips 373

ASE EDUCATION FOUNDATION CORRELATION

IMMR TASK	TST TASK	MTST TASK	PRIORITY	JOB SHEET NUMBER
I. DIESEL ENGINES				
I.A.1	I.A.1	I.A.1	P-1	1.3, 4.1, 4.2, 4.3, 4.4, 4.5, 4.6, 4.7, 4.8, 4.9, 4.11, 4.12
I.A.2	I.A.2	I.A.2	P-1	4.4, 4.5, 4.6, 4.7
I.A.4	I.A.4	I.A.4	P-2	2.1, 4.1, 4.2
I.D.2	I.D.2	I.D.2	P-1	2.1, 4.1, 4.2, 4.3, 4.4, 4.5, 4.6, 4.7, 4.8, 4.9, 4.11
I.D.3	I.D.3	I.D.3	P-1	4.4, 4.5, 4.7
I.E.1	I.E.1	I.E.1	P-1	4.1, 4.2, 4.3, 4.4, 4.5, 4.6, 4.7, 4.8, 4.9, 4.11, 4.12
I.E.2	I.E.2	I.E.2	P-1	4.4, 4.5, 4.6, 4.7, 4.12
I.E.3	I.E.3	I.E.3	P-1	4.12
I.E.4	I.E.4	I.E.4	P-1	4.12
I.E.5	I.E.5	I.E.5	P-1	4.12
I.E.6	I.E.6	I.E.6	P-1	4.12
I.E.7	I.E.7	I.E.7	P-1	4.12
I.E.8	I.E.8	I.E.8	P-1	4.12
I.F.2	I.F.2	I.F.2	P-1	4.1, 4-2, 4.3, 4.4, 4.5, 4.6, 4.7, 4.8, 4.9, 4.11, 4.12
I.G.1	I.G.1	I.G.1	P-1	4.1, 4.2
I.G.2	I.G.2	I.G.2	P-1	4.1, 4.2, 4.3, 4.9, 4.11
I.G.4	I.G.4	I.G.4	P-1	4.2, 4.3, 4.9, 4.11
II. DRIVE TRAIN				
II.A.1	II.A.1	II.A.1	P-1	1.3, 20.1, 21.1, 21.2, 21.3
II.A.2	II.A.2	II.A.2	P-1	4.2
II.B.1	II.B.1	II.B.1	P-1	4.2, 14.1, 14.2
II.B.2	II.B.2	II.B.2	P-1	4.2, 4.3, 4.9, 4.11, 14.2
	II.B.3	II.B.3	P-2	14.2
	II.B.4	II.B.4	P-1	14.1, 14.2
	II.B.5	II.B.5	P-1	14.2
	II.B.6	II.B.6	P-1	14.1, 14.2
	II.B.7	II.B.7	P-1	14.1, 14.2
	II.B.8	II.B.8	P-1	14.1, 14.2
	II.B.9	II.B.9	P-1	14.2
II.C.1	II.C.1	II.C.1	P-1	4.5, 16.1, 19.2
II.C.2	II.C.2	II.C.2	P-1	16.1, 19.1, 19.2
II.C.3	II.C.3	II.C.3	P-1	16.1, 19.1
II.C.4	II.C.4	II.C.4	P-1	4.2, 4.3, 4.8, 4.9, 4.11, 16.1, 19.1, 19.2
II.C.5	II.C.5	II.C.5	P-2	4.2, 4.3, 4.9, 4.11, 16.1, 19.1
II.C.6	II.C.6	II.C.6	P-2	16.1
II.C.7	II.C.7	II.C.7	P-1	16.1
	II.C.8	II.C.8	P-2	15.1, 16.1
	II.C.9	II.C.9	P-1	19.2

IMMR TASK	TST TASK	MTST TASK	PRIORITY	JOB SHEET NUMBER
	II.C.10	II.C.10	P-2	16.1
	II.C.11	II.C.11	P-2	15.1, 16.1, 17.1, 18.1
	II.C.12	II.C.12	P-3	16.1, 19.1
	II.C.13	II.C.13	P-3	16.1
	II.C.14	II.C.14	P-2	16.1, 19.2
	II.C.15	II.C.15	P-2	17.1, 18.1, 19.1, 19.2, 21.1, 21.2, 21.3
	II.C.16	II.C.16	P-2	20.1
II.D.1	II.D.1	II.D.1	P-1	4.2, 4.5, 4.6, 4.7, 4.11, 22.1
	II.D.3	II.D.3	P-1	22.1
II.E.1	II.E.1	II.E.1	P-1	4.2, 4.3, 4.8, 4.9, 4.11, 24.2
II.E.2	II.E.2	II.E.2	P-1	4.2, 4.3, 4.5, 4.6, 4.7, 4.8, 4.9, 4.11, 24.1, 24.2
II.E.3	II.E.3	II.E.3	P-2	24.1, 24.2
II.E.4	II.E.4	II.E.4	P-2	23.1, 24.2
II.E.5	II.E.5	II.E.5	P-1	23.1, 24.2
	II.E.6	II.E.6	P-3	24.1, 24.2
	II.E.9	II.E.9	P-2	23.1, 24.1, 24.2
III. BRAKES				
III.A.1	III.A.1	III.A.1	P-1	1.3, 4.2, 4.10, 28.1, 28.2, 29.1, 30.1, 31.1, 31.2, 31.3, 31.4
III.A.2	III.A.2	III.A.2	P-1	4.2, 4.3, 4.9, 28.1, 28.2, 29.1, 30.1, 31.1, 31.2, 31.3, 31.4
III.A.3	III.A.3	III.A.3	P-1	29.1, 31.1, 31.2
	III.A.4	III.A.4	P-1	29.1, 31.1, 31.2
III.B.1	III.B.1	III.B.1	P-1	4.2, 31.1, 31.2
III.B.2	III.B.2	III.B.2	P-1	4.2, 31.1, 31.2
	III.B.3	III.B.3	P-1	28.2, 31.1, 31.2, 31.3, 31.4
	III.B.4	III.B.4	P-3	31.1, 31.2
	III.B.5	III.B.5	P-1	31.1, 31.2
	III.B.6	III.B.6	P-1	31.1, 31.2
	III.B.7	III.B.7	P-1	31.1, 31.2
	III.B.8	III.B.8	P-1	31.1, 31.2
III.C.1	III.C.1	III.C.1	P-1	4.2, 4.10, 31.1, 31.2
III.C.2	III.C.2	III.C.2	P-1	4.2, 4.10, 28.1, 28.2, 31.1, 31.2
III.C.3	III.C.3	III.C.3	P-1	4.2, 4.10, 28.2, 31.1, 31.2
III.C.4	III.C.4	III.C.4	P-1	3.1, 4.2, 31.1, 31.2
III.C.5	III.C.5	III.C.5	P-1	3.1, 4.2, 31.1, 31.2
III.C.6	III.C.6	III.C.6	P-1	3.1, 4.2, 4.10, 28.2, 31.1, 31.2
	III.C.7	III.C.7	P-1	28.2, 31.1, 31.2
III.D.1	III.D.1	III.D.1	P-1	4.2, 4.3, 4.8, 4.9, 4.10, 28.2, 31.1, 31.2
III.D.2	III.D.2	III.D.2	P-1	4.10, 28.2, 31.1, 31.2
III.D.3	III.D.3	III.D.3	P-1	4.10, 28.2, 31.1, 31.2

IMMR TASK	TST TASK	MTST TASK	PRIORITY	JOB SHEET NUMBER
III.D.4	III.D.4	III.D.4	P-1	4.10, 31.1, 31.2
	III.D.5	III.D.5	P-2	28.2, 31.1, 31.2
III.E.1	III.E.1	III.E.1	P-1	4.2, 4.3, 29.1
III.E.2	III.E.2	III.E.2	P-1	4.2, 4.3, 4.9, 29.1
III.E.3	III.E.3	III.E.3	P-1	4.2, 29.1
	III.E.4	III.E.4	P-2	29.1
	III.E.5	III.E.5	P-2	29.1
	III.E.6	III.E.6	P-3	29.1
	III.E.7	III.E.7	P-2	29.1
	III.E.8	III.E.8	P-2	29.1
III.F.1	III.F.1	III.F.1	P-1	28.2, 29.1, 31.1, 31.2
III.F.2	III.F.2	III.F.2	P-1	28.2, 29.1, 31.1, 31.2
III.F.3	III.F.3	III.F.3	P-1	28.2, 29.1, 31.1, 31.2
III.G.1	III.G.1	III.G.1	P-1	2.1, 4.2
III.H.1	III.H.1	III.H.1	P-1	28.2, 29.1, 31.1, 31.2
III.I.1	III.I.1	III.I.1	P-1	4.2, 28.2, 29.1, 30.1, 31.1, 31.2
III.I.2	III.I.2	III.I.2	P-2	4.2, 28.2, 29.1, 30.1, 31.1, 31.2
	III.I.3	III.I.3	P-2	29.1, 30.1, 31.1, 31.2
	III.I.5	III.I.5	P-1	29.1, 31.2
	III.I.6	III.I.6	P-1	29.1, 30.1, 31.1, 31.2
	III.I.7	III.I.7	P-2	29.1, 30.1, 31.2
	III.I.8	III.I.8	P-3	28.2, 29.1, 31.1, 31.2
III.J.1	III.J.1	III.J.1	P-1	4.2, 4.3, 4.8, 4.9, 4.10, 4.11, 27.4, 29.1, 31.1, 31.2
III.J.2	III.J.2	III.J.2	P-2	4.2, 4.10, 27.4, 29.1, 31.1, 31.2
IV. SUSPENSION AND STEERING				
IV.A.1	IV.A.1	IV.A.1	P-1	1.3, 25.1, 25.2, 25.3, 25.4, 25.5, 25.6, 26.1, 26.2, 26.3, 27.1, 27.2, 27.3, 27.4, 32.1, 33.1, 33.2
IV.B.1	IV.B.1	IV.B.1	P-1	4.2
IV.C.1	IV.C.1	IV.C.1	P-1	4.2, 4.3, 4.8, 4.9, 4.11, 25.4, 25.5, 25.6
IV.C.2	IV.C.2	IV.C.2	P-2, P-1, P-1	4.2, 25.4
	IV.C.3	IV.C.3	P-1	25.4, 25.5, 25.6
	IV.C.4	IV.C.4	P-2	25.4, 25.5, 25.6
	IV.C.6	IV.C.6	P-2	25.4
IV.D.1	IV.D.1	IV.D.1	P-1	4.2
IV.E.1	IV.E.1	IV.E.1	P-1	4.2, 4.10, 26.1, 26.2, 26.3
IV.E.2	IV.E.2	IV.E.2	P-1	4.2, 4.10, 26.1, 33.1
IV.E.3	IV.E.3	IV.E.3	P-1	26.1, 26.2, 26.3
IV.E.4	IV.E.4	IV.E.4	P-3	26.1
IV.E.5	IV.E.5	IV.E.5	P-1	4.2, 4.10, 26.2, 26.3
IV.E.6	IV.E.6	IV.E.6	P-1	4.2, 26.2
	IV.E.8	IV.E.8	P-1	26.1, 26.2, 26.3

IMMR TASK	TST TASK	MTST TASK	PRIORITY	JOB SHEET NUMBER
IV.F.1	IV.F.1	IV.F.1	P-3, P-1, P-1	25.1, 25.2, 25.3
	IV.F.3	IV.F.3	P-2	4.2
	IV.F.4	IV.F.4	P-2	4.2, 25.1, 25.2
	IV.F.5	IV.F.5	P-1	4.2, 25.3
	IV.F.6	IV.F.6	P-2	4.2
	IV.F.7	IV.F.7	P-3	4.2, 25.5, 25.6
	IV.F.8	IV.F.8	P-2	4.2, 25.3
IV.G.1	IV.G.1	IV.G.1	P-1	4.2, 4.3, 4.8, 4.9, 4.10
IV.G.2	IV.G.2	IV.G.2	P-2	27.4
IV.G.3	IV.G.3	IV.G.3	P-1	4.2, 4.3, 4.8, 4.9, 4.10, 4.11, 27.1, 27.2, 27.3
IV.H.1	IV.H.1	IV.H.1	P-1	4.2, 4.10, 4.13, 34.1, 34.2, 34.3
IV.H.2	IV.H.2	IV.H.2	P-1	4.10, 4.13, 32.1, 33.1, 33.2, 34.2, 34.3
IV.H.3	IV.H.3	IV.H.3	P-3	4.13, 32.1, 33.2, 34.1, 34.2, 34.3
IV.H.4	IV.H.4	IV.H.4	P-1, P-2, P-2	4.10, 4.13
V. ELECTRICAL/ELECTRONIC SYSTEMS				
V.A.1	V.A.1	V.A.1	P-1	5.1, 5.2, 6.1, 7.1, 8.1, 8.2, 9.1, 9.2, 9.3, 10.1, 10.2, 10.3, 11.1, 11.3, 12.1, 12.2, 12.3
V.A.2	V.A.2	V.A.2	P-1	5.1, 5.2, 6.2, 7.1, 8.1, 8.2, 9.1, 9.2, 9.3, 10.1, 10.2, 10.3, 11.1, 11.2, 11.3
V.A.3	V.A.3	V.A.3	P-1	4.2, 5.1, 5.2, 6.1, 6.2, 7.1, 8.1, 8.2, 9.1, 9.2, 9.3, 10.3, 11.1, 11.2, 11.3
V.A.4	V.A.4	V.A.4	P-1	5.1, 5.2, 6.1, 6.2, 7.1, 8.1, 8.2, 9.1, 9.2, 11.1, 11.2, 11.3
V.A.5	V.A.5	V.A.5	P-1	4.2, 5.1, 5.2, 6.1, 6.2, 7.1, 8.1 8.2, 9.1, 9.2, 11.1, 11.3, 12.1
V.A.6	V.A.6	V.A.6	P-1	5.2
V.A.7	V.A.7	V.A.7	P-1	5.2, 9.1
V.A.8	V.A.8	V.A.8	P-1	5.2, 9.1, 9.3
V.A.9	V.A.9	V.A.9	P-2, P-2, P-1	4.2, 5.2, 6.1, 6.2, 7.1, 8.1, 8.2, 9.1, 9.2, 9.3, 11.1, 12.1, 12.3
V.A.10	V.A.10	V.A.10	P-2	9.2, 12.1, 12.2, 12.3
V.A.11	V.A.11	V.A.11	P-1	5.1, 6.1, 6.2, 8.1, 8.2, 11.1, 11.2, 12.1, 12.2, 12.3
V.B.1	V.B.1	V.B.1	P-1	4.2, 5.2, 7.1, 9.1, 9.3, 10.3, 11.3
V.B.2	V.B.2	V.B.2	P-1	4.2, 4.3, 4.7, 4.9, 4.11, 5.2, 7.1, 8.1, 8.2, 9.1, 9.3, 10.3, 11.3
V.B.3	V.B.3	V.B.3	P-1	4.2, 4.3, 4.7, 4.9, 5.2, 7.1, 8.2, 9.1, 9.3, 10.3
V.B.4	V.B.4	V.B.4	P-1	4.2, 8.1, 8.2, 9.3, 10.3
V.B.5	V.B.5	V.B.5	P-1	10.3
V.B.6	V.B.6	V.B.6	P-2, P-2, P-1	7.1, 9.3, 10.3
V.C.1	V.C.1	V.C.1	P-1	5.2, 9.1, 9.2, 9.3, 10.3, 11.3
V.C.2	V.C.2	V.C.2	P-1	4.2, 5.2, 9.1, 9.2, 9.3, 10.2, 10.3, 11.3
V.C.3	V.C.3	V.C.3	P-1	4.2, 5.2, 9.1, 9.3, 10.1, 10.3, 11.3
	V.C.5	V.C.5	P-3, P-2	4.2

IMMR TASK	TST TASK	MTST TASK	PRIORITY	JOB SHEET NUMBER
V.D.1	V.D.1	V.D.1	P-1	5.2, 8.1, 8.2, 9.3, 10.3, 11.3
V.D.2	V.D.2	V.D.2	P-1	5.2, 8.1, 8.2, 9.3, 10.3
V.D.3	V.D.3	V.D.3	P-1	8.1, 8.2, 9.3, 10.3
V.D.4	V.D.4	V.D.4	P-1	5.2, 8.1, 8.2, 9.3, 10.3
V.D.5	V.D.5	V.D.5	P-1	4.2, 5.2, 8.1, 8.2, 9.3, 10.3, 11.3
	V.D.6	V.D.6	P-1	4.2
V.E.1	V.E.1	V.E.1	P-1	4.7, 8.2
V.E.2	V.E.2	V.E.2	P-1	4.2, 4.3, 4.7, 4.8, 4.9, 4.11
V.E.4	V.E.4	V.E.4	P-1, P-1, P-2	
V.F.1	V.F.1	V.F.1	P-1	4.2
V.F.2	V.F.2	V.F.2	P-2	6.1, 11.1, 11.2, 12.1
VI. HEATING, VENTILATION, AND AIR CONDITIONING (HVAC)				
VI.A.1	VI.A.1	VI.A.1	P-1	4.14, 35.1, 35.2
VI.A.2	VI.A.2	VI.A.2	P-1	4.2, 4.14, 35.1, 35.2
VI.A.3	VI.A.3	VI.A.3	P-1	4.4, 4.14, 35.1, 35.2
	VI.A.6	VI.A.6	P-1	4.2
VI.B.1	VI.B.1	VI.B.1	P-1	4.4, 4.14, 35.2
VI.B.2	VI.B.2	VI.B.2	P-1	4.4, 4.14, 35.2
VI.B.3	VI.B.3	VI.B.3	P-1	4.14, 35.2
VI.C.1	VI.C.1	VI.C.1	P-1	4.14, 35.2
VI.C.2	VI.C.2	VI.C.2	P-1	4.14, 35.1, 35.2
VI.C.3	VI.C.3	VI.C.3	P-2, P-2, P-1	4.14, 35.2
VI.D.1	VI.D.1	VI.D.1	P-1	4.4, 4.5, 4.12, 4.14, 35.2
VII. CAB				
VII.A.1	VII.A.1	VII.A.1	P-1	1.2, 1.3
VII.B.1	VII.B.1	VII.B.1	P-1	2.1
VII.B.2	VII.B.2	VII.B.2	P-1	2.1, 4.2, 4.3, 4.8, 4.9, 4.10, 4.11
VII.D.1	VII.D.1	VII.D.1	P-1	4.1, 4.2, 4.3, 4.9, 4.11
VIII. HYDRAULICS				
VIII.A.1	VIII.A.1	VIII.A.1	P-3	13.1
VIII.A.3	VIII.A.3	VIII.A.3	P-3	13.1
VIII.A.4	VIII.A.4	VIII.A.4	P-3	13.1
		VIII.A.8	P-3	13.1
		VIII.B.2	P-3	13.1
		VIII.B.4	P-3	13.1
		VIII.C.1	P-3	13.1
		VIII.D.2	P-3	13.1
		VIII.D.3	P-3	13.1
		VIII.E.1	P-3	13.1
		VIII.E.2	P-3	13.1

Preface

The Student Workbook has been expanded for the seventh edition of *Heavy-Duty Truck Systems*, to accommodate some of the new technology introduced. The primary objective of the Student Workbook is to reinforce some of the most basic everyday service facility tasks and correlate each to the latest ASE Education Foundation competencies. Networking skills are integrated into the conclusion of each set of tasks, reflecting the requirement for truck technicians to make use of online research and reference sources. Most of the end of chapter network tasks are simple to perform and designed to familiarize student with how truck repair technology uses the Internet.

In keeping with the objective of reinforcing performance skills, most of the exercises in the Student Workbook attempt to make the connection between the theoretical concepts in the core textbook and the hands-on application of that knowledge. A truck technician is first required to understand a technical concept, and then to make that critical connection between a concept and the hands-on skills of diagnosis and repair. The Student Workbook attempts to familiarize students with everyday shop floor practice.

MAINTENANCE

Maintenance practices continue to be a feature of this edition of the Student Workbook because so much routine maintenance in the trucking industry is performed by entry-level technicians. The scope of maintenance practices includes condition-based maintenance (CBM) along with more traditional preventive maintenance (PM). The task sheets in Chapter 4 are based on different OEM-recommended maintenance practices. The terms used to describe maintenance procedures, such as *B-inspection*, *winterization-inspection*, *chassis-lube inspection*, and others, are used throughout the industry. Most fleets adapt these OEM-recommended practices to suit their own specific needs and in many cases they can be more exacting than the original OEM checklists. Although some may regard the performing of a maintenance inspection service on vehicles as being tedious, major fleets regard their PM practices as a key to avoid unscheduled vehicle downtime. In addition, they value technicians with an eye for detail who can identify a potential failure before it results in a breakdown. If you are targeting employment with transport fleets, you should attempt to complete all of the checklist exercises in Chapter 4 of the Student Workbook, making sure that you know exactly what should be done when you cannot check off a category as being okay.

LEARNING OBJECTIVES

Each chapter in the Student Workbook begins by listing the learning objectives of each textbook chapter exactly as they appear in the core text. This is followed by a few practice questions: The objective here is to test your knowledge in some key subject areas addressed in the chapter. The learning objectives in each chapter should be used in conjunction with the Summary review that appears at the end of each chapter in the textbook. Assuming that a student understands the material covered by a chapter, studying the Summary review bullets and end-of-chapter questions is a great way of preparing for a test.

JOB SHEETS

Between one and fourteen job sheets are provided with each chapter. These identify some typical shop or service garage procedures that connect with the subject matter of the chapter. In some cases, you will need specialized equipment and tools to perform these job sheets. However, in cases where you cannot access the tools or equipment identified in the job sheet, you should still be able to get some benefit from the tasks simply by following the procedure through using the text.

JOB SHEET EVALUATIONS

At the end of each job sheet is a pair of evaluation tables. The first is a self-evaluation, and this is probably the most important of the pair. Research has continually indicated that students evaluate their own performance more critically than either teachers or peers.

Be honest when completing a self-evaluation. It can help you identify areas in which you require additional study or hands-on experience. The instructor evaluation can also be valuable. It may help you identify key areas of your work performance that can be improved. Furthermore, it is important because the grade you earn for a course usually depends on how an instructor rates your performance. Knowing how supervisors evaluate your work habits is a critical aspect to succeeding in the workplace.

ONLINE TASKS

Some online tasks are provided for each chapter. These online tasks are designed to get you accustomed to using Internet search engines and building a useful inventory of references in your bookmark/favorites folders. They are by no means intended to be comprehensive. All students are reminded that although the Internet is an outstanding research resource, it is not infallible. That said, you should have no problem exceeding the objectives of the online tasks, and it is important to remember that the scope of the Internet vastly exceeds Wikipedia: Make a habit of relying on more than just one or two information resources.

KEY POINTS

Finally, each short chapter in the Workbook finishes with some study tips. You create these by identifying what you think are five key points in each chapter. You will find it easiest to complete this section if you do so immediately after studying each chapter. The best way to complete the key points is to begin each with a one-word bullet and then add to this as few words as possible, just enough to remind yourself of the concept later on. An example of how to do this would be as follows:

Key point: linehaul—a truck used in terminal-to-terminal operation
Key point: vocational—construction/fire trucks/mixers/packers

Get the idea? Try it. The one-word bullet is explained in point form immediately after. This can be an effective method of self-study.

USING THE STUDENT WORKBOOK EFFECTIVELY

The key elements in each chapter of the Student Workbook are organized in the following sequence:

- Learning Objectives
- Practice Questions
- Job Sheets
- Online Tasks
- Study Tips

Ideally, each set of exercises in the Student Workbook should be completed as you finish studying the chapter in the textbook. The job sheets can be an especially useful study reference. Because they require you to actually perform hands-on activities, they are often more likely to stick in your mind. Most truck technicians learn best by doing rather than by reading. By taking a look at a job sheet you may have completed a year ago just before taking a test, you should be able to instantly recall the procedure.

ASE EDUCATION FOUNDATION TASK REQUIREMENTS

Inspection, Maintenance and Minor Repair (IMMR)	**540 hours** combined classroom and lab/shop instructional activities
Truck Service Technology (TST)	**740 hours** combined classroom and lab/shop instructional activities

Master Truck Service Technology (MTST)   **1040 hours**
combined classroom and lab/shop instructional activities

Accreditation Level	Tasks/Hours
M/H Truck Inspection, Maintenance, and Minor Repair (IMMR)	• 199 Tasks • 540 hours
M/H Truck Service Technology (TST)	• 329 Tasks • 740 hours
Master M/H Truck Service Technology (MTST)	• 399 Tasks • 1040 hours

The acronyms IMMR, TST, and MTST will be used in the worksheets throughout the Workbook.

1 Introduction to Servicing Heavy-Duty Trucks

Objectives

After reading this chapter, you should be able to:

- Explain the basic truck classifications.
- Define gross vehicle weight (GVW).
- Identify the major original equipment manufacturers (OEMs).
- Classify a truck by the number of axles it has.
- Identify an on-highway truck's major systems and its related components.
- Identify various career opportunities in the heavy-duty trucking industry.
- Explain the job classifications offered by the truck industry to qualified and experienced technicians.
- Understand the National Institute for Automotive Service Excellence (ASE) certification program and how it benefits technicians.
- Distinguish ASE T-, S-, and H-qualifications and understand what is required for Master Technician status.
- Identify some methods of maintaining currency as a truck technician.
- Explain how the electronic logging device (ELD) mandate affects how technicians work on vehicles.

PRACTICE QUESTIONS

1. If a truck is classified as a 6 × 2, what does the 2 indicate?
   a. total number of axles
   b. steering axle wheels
   c. driven wheels
   d. driven axles

2. Which of the following designations would represent the highest weight category for on-highway trucks in North America?
   a. Class 4
   b. Class 7
   c. Class 8
   d. Class 10

3. Which of the following terms best describes how a linehaul truck is used?
   a. terminal to terminal
   b. pickup and delivery
   c. aggregate haulage
   d. courier dropoffs

4. Which of the following would be described as vocational trucks?
   a. garbage packers
   b. dump trucks
   c. fire trucks
   d. all of the above

5. Which ASE certification category covers truck air brake components?
    a. T2
    b. T4
    c. T6
    d. T8
6. Which of the following shops would usually deal with a truck warranty complaint?
    a. specialty service shop
    b. fleet service shop
    c. independent service shop
    d. OEM dealership
7. Which of the following is used to couple a tractor to a semi-trailer?
    a. bar hitch
    b. fifth wheel
    c. pintle hook
    d. ball hitch
8. Which type of wheel would not be found on a heavy-duty highway truck?
    a. cast spoke
    b. aluminum disc
    c. steel disc
    d. extruded spoke
9. Research and write down as many manufacturers of Class 6, 7, and 8 trucks as you can.

10. Research and write down as many manufacturers of trailers as you can.

JOB SHEET 1.1

Name _____ Station _____ Date _____

Build a Truck OEM Folder in Your Bookmark or Favorites Log, and Load with all the Major Truck OEMs.

Performance Objective(s): Use an Internet search engine to create a reference log consisting of all the major truck chassis OEMs.

ASE Education Foundation Correlation

This job sheet addresses the following ASE Education Foundation task(s):

- Narrative for language arts-related academic skills
- Workplace Skills: Sections D and H

Tools and Materials: A personal computer with Web browser software and a Web connection

Protective Clothing: None required

PROCEDURE

1. Create a folder in your Bookmark or Favorites portal and load it with the Web page of each major truck OEM: Daimler Trucks North America (Freightliner, Western Star), Navistar-International Trucks, Volvo/Mack Trucks, DAF-Paccar (Kenworth, Peterbilt), Tesla, Nikola, Caterpillar, Euclid, Fuso, Isuzu, and Hino. Use the search engine to locate their corporate home page.

   Task completed _____

2. Select a search engine to locate some of the major component suppliers to truck OEMs and select a handful to profile in your Bookmark or Favorites folder. Some examples to get you started are Eaton Fuller Roadranger, Dana (Spicer), Alcoa, Allison Transmission, Voith, Bendix, Hendrickson, Holland Suspensions (Neway), TRW Automotive (Ross), Meritor-WABCO, Haldex, Accuride, Carrier Refrigeration, Chicago Rawhide, Bridgestone (Firestone), Michelin, Fontaine, Jost, SAF-Holland, Remy International, Nexiq, Motorola, Delphi, Wabash, Great Dane, Heil, Trail King, Stroughton, Trailmobile, etc.

   Task completed _____

3. Select a couple of the manufacturer websites you have logged in your Bookmark/Favorites/folder: Make a note of the objective of each web page and determine whether you think it enhances their product or corporate image.

   Task completed _____

4. Study the two columns of truck OEMs below. Each OEM in the left column has common ownership with an OEM on the right. Connect by drawing a line between the two.

Volvo	Kenworth
Western Star	Mack Trucks
Peterbilt	Freightliner

   Task completed _____

5. Some of manufacturers supplying the trucking industry provide excellent customer support in the form of free service literature and diagnostic software. See if you can identify at least half a dozen of these companies and create a diagnostic reference folder. A clue to get you going: Bendix.

   Task completed _____

STUDENT SELF-EVALUATION

Check	Level	Competency	Comments
	4	Mastered task	
	3	Competent but need further help	
	2	Needed a lot of help	
	0	Did not understand the task	

INSTRUCTOR EVALUATION

Check	Level	Competency	Comments
	4	Mastered task	
	3	Competent but needs further help	
	2	Requires more training	
	0	Unable to perform task	

JOB SHEET 1.2

Name _____ Station _____ Date _____

Use the Internet to Research Information.

Performance Objective(s): Use the Internet to research truck technical and OEM product information and print the results of the search into a portfolio.

ASE Education Foundation Correlation

This job sheet addresses the following ASE Education Foundation task(s):

VII. Cab
A. General
1. Research vehicle service information including, vehicle service history, service precautions, and technical service bulletins. (IMMR, TST, MTST) P-1.

- Narrative for language arts-related academic skills
- Workplace skills: Sections D and K
- Narrative for Science Related Academic Skills

Tools and Materials: A personal computer with Web browser software and Web access

Protective Clothing: None required

PROCEDURE

1. Log on to the Internet.

   Task completed _____

2. Use a search engine or your Favorites/Bookmark file to log information on the following organizations:

ASE	Task completed _____
ATA-TMC	Task completed _____
SAE	Task completed _____

   Ask yourself these questions:

What role does each of the preceding organizations play?	Task completed _____
What kind of influence could each organization have on my career?	Task completed _____

3. Use a search engine or your Favorites/Bookmark log and select one major truck OEM. List the entire product line of the OEM you selected. Try to correlate the OEM chassis coding to numeric class categories. Example: A Freightliner Cascadia chassis falls into the Class 8 category. Attempt to identify OEM specialties such as linehaul, vocational, pickup and delivery, etc.

   Task completed _____

STUDENT SELF-EVALUATION

Check	Level	Competency	Comments
	4	Mastered task	
	3	Competent but need further help	
	2	Needed a lot of help	
	0	Did not understand the task	

INSTRUCTOR EVALUATION

Check	Level	Competency	Comments
	4	Mastered task	
	3	Competent but needs further help	
	2	Requires more training	
	0	Unable to perform task	

JOB SHEET 1.3

Name _____ Station _____ Date _____

Walk-Around Inspection of a Commercial Vehicle

Performance Objective(s): Perform a walk-around inspection of a commercial vehicle and record all the visible data that defines its intended usage.

ASE Education Foundation Correlation

This job sheet addresses the following ASE Education Foundation task(s):

I. **Diesel Engines**
   A. **General**
   1. Research vehicle service information, including fluid type, vehicle service history, service precautions, and technical service bulletins. (IMMR, TST, MTST) P-1.

II. **Drive Train**
   A. **General**
   1. Research vehicle service information, including fluid type, vehicle service history, service precautions, and technical service bulletins. (IMMR, TST, MTST) P-1.

III. **Brakes**
   A. **General**
   1. Research vehicle service information, including fluid type, vehicle service history, service precautions, and technical service bulletins. (IMMR, TST, MTST) P-1.

IV. **Suspension & Steering Systems.**
   A. **General**
   1. Research vehicle service information, including fluid type, vehicle service history, service precautions, and technical service bulletins. (IMMR, TST, MTST) P-1.

VII. **Cab**
   A. **General**
   1. Research vehicle service information, including fluid type, vehicle service history, service precautions, and technical service bulletins. (IMMR, TST, MTST) P-1.

- Narrative for language arts-related academic skills
- Workplace Skills: Sections D and H

Tools and Materials: A commercial vehicle (Class 4 to 8), school bus, transit bus, or offroad construction vehicle.

Protective Clothing: Standard PPE: coveralls, safety footwear, eye protection, and mechanic's gloves.

PROCEDURE:

1. Record the serial number _____

   Task completed _____

2. Identify the manufacturer, model, and year of manufacture: _____

   Task completed _____

3. Identify the axle configuration: ....X....

   Task completed _____

   Circle the vehicle class designation: Class 4 5 6 7 8

   Task completed _____

4. Engine OEM _____

   Engine power and torque ratings _____ BHP _____ lb./ft

   Transmission manufacturer and ratios _____

   Final drive manufacturer and ratio(s) _____

   Circle foundation brake type: S-cam   Disc   Air   Air-over-hydraulic   Hydraulic

   Identify 5th wheel manufacturer and model _____

   Identify wheel type: _____

   Identify tire sizes, manufacturer, and recommended inflation pressures _____

   Task completed _____

5. Circle the intended vehicle application:

   Linehaul Pick-up and Delivery Vocational People mover Construction

   Task completed _____

6. Identify the vehicle ownership:

   Fleet   Owner-operator   Lease

   Task completed _____

STUDENT SELF-EVALUATION

Check	Level	Competency	Comments
	4	Mastered task	
	3	Competent but need further help	
	2	Needed a lot of help	
	0	Did not understand the task	

INSTRUCTOR EVALUATION

Check	Level	Competency	Comments
	4	Mastered task	
	3	Competent but needs further help	
	2	Requires more training	
	0	Unable to perform task	

STUDY TIPS

Identify 5 key points in Chapter 1. Try to be as brief as possible.

Key point 1 _____

Key point 2 _____

Key point 3 _____

Key point 4 _____

Key point 5 _____

2 Shop Safety and Operations

Objectives

After reading this chapter, you should be able to:

- Explain the special notations in the text labeled SHOP TALK, CAUTION, and WARNING.
- Identify the basic procedures for lifting and carrying heavy objects and materials.
- Explain how to use personal protective equipment (PPE).
- Identify the UL requirements of shop safety boots and ESR footwear.
- Describe safety warnings as they relate to work area safety.
- Identify the different classifications of fires and the proper procedures for extinguishing each.
- Operate the various types of fire extinguishers based on the type of extinguishing agent each uses.
- Identify the four categories of hazardous waste and their respective hazards to health and the environment.
- Explain laws regulating hazardous materials, including both the "Right-to-Know" and employee/employer obligations.
- Identify which types of records are required by law to be maintained on trucks involved in interstate shipping.
- Outline the precautions required to work on hybrid hydraulic, hybrid electric, and gaseous fueled vehicles.
- Explain what hydraulic pinhole injection is and the action required if you suspect it.
- List the safety requirements required to work around high-voltage electrical equipment.
- Identify the precautions required to work safely with oxyacetylene equipment.
- Identify the precautions required to work safely with electric welding stations.
- Discuss the role of computers in the administration, logistics, and maintenance management of transport truck operations.
- Understand the lockout/tagout (LOTO) policy and the legal consequences of ignoring it.
- Identify the types of LOTO mechanisms used in a truck shop and explain the importance of using them.

PRACTICE QUESTIONS

1. Make a list of clothing and personal protection equipment (PPE) that should be worn by a technician in a truck shop.

2. Which of the following is most essential when performing flame-cutting and welding operations?
   a. safety boots
   b. welding gauntlets
   c. welding helmet/goggles with correct filter lens for the procedure
   d. all of the above

3. Research which lens filter grades are acceptable for use when performing oxyacetylene flame cutting.

4. Research which lens filter grades are acceptable for use when performing arc welding procedures.

5. When diesel fuel ignites, which of the following should **never** be done when attempting to extinguish the fire?
   a. use a Class-B-rated fire extinguisher
   b. douse with water
   c. smother with foam
   d. use a carbon dioxide fire extinguisher

6. Which of the following types of fire extinguishers should be used to extinguish a fire that has ignited inside an electric arc welding unit?
   a. water
   b. foam
   c. carbon dioxide
   d. dry chemical

7. Who administers Right-to-Know legislation?
   a. The American Trucking Associations (ATA)
   b. The Technology and Maintenance Council (TMC)
   c. The Society of Automotive Engineers (SAE)
   d. Occupational Safety and Health Administration (OSHA)

8. A material that can dissolve metals and burn skin is:
   a. corrosive
   b. reactive
   c. dangerous
   d. all of the above

9. Look up the term *toxin* in a dictionary. Which of the following words best matches it?
   a. poison
   b. acid
   c. explosive
   d. radioactive

10. Locate any liquid substance in a container in your shop, such as a can of paint or solvent. Obtain the material safety data sheet (MSDS). Check out the reactivity, flammability, and toxicity fields and make a list of the first-aid procedures required.

JOB SHEET 2.1

Name _____ Station _____ Date _____

Check out the National Highway Traffic Safety Administration (NHTSA) Web site.

Performance Objective(s): Retrieve information from the NHTSA Web site.

ASE Education Foundation Correlation

This job sheet addresses the following ASE Education Foundation task(s):

- Narrative for language arts-related academic skills
- Workplace skills: Sections D, F, H, and I

Tools and Materials: A personal computer with Web browser software and Web access

Protective Clothing: None required

PROCEDURE

1. Log on to the Internet. Use the touchpad or keypad to enter http://www.nhtsa.gov in the address field and enter the site.

   Task completed _____

2. In the NHTSA Web site, navigate to the Safety Problems and Issues section.

   Task completed _____

3. Download and print out one of the articles in the Safety Problems and Issues section. Discuss how you would resolve the problem. Note how the OEM concerned plans to resolve the problem.

   Task completed _____

4. Write down some of the benefits of having the NHTSA administered by the federal government. How might this be different if the NHTSA were administered by industry?

   Task completed _____

STUDENT SELF-EVALUATION

Check	Level	Competency	Comments
	4	Mastered task	
	3	Competent but need further help	
	2	Needed a lot of help	
	0	Did not understand the task	

INSTRUCTOR EVALUATION

Check	Level	Competency	Comments
	4	Mastered task	
	3	Competent but needs further help	
	2	Requires more training	
	0	Unable to perform task	

JOB SHEET 2.2

Name _____ Station _____ Date _____

Safely Start and Operate a Heavy-Duty Truck.

Performance Objective(s): Start up and drive a medium- or heavy-duty truck from a yard and safely park it in a shop bay. This exercise assumes that the student is already qualified to drive an automobile.

ASE Education Foundation Correlation

This job sheet addresses the following ASE Education Foundation task(s):

I. Diesel Engines
  A. General
  4. Check engine operation (starting and running) including: noise, vibration, smoke, etc.; determine needed action. (IMMR, TST) P-2.
  4. Diagnose engine operation (starting and running) including: noise, vibration, smoke, etc.; determine needed action. (MTST) P-2.

  D. Lubrication Systems
  2. Check engine oil level, condition, and consumption; take engine oil sample; determine needed action. (IMMR, TST, MTST) P-1.

III. Brakes
  G. Hydraulic Brakes: Parking Brake System.
  1. Check parking brake operation; inspect parking brake application and holding devices. (IMMR) P-1.
  1. Check parking brake operation; inspect parking brake application and holding devices; adjust, repair, and/or replace as needed. (TST, MTST) P-1.

VII. Cab
  B. Instruments and Controls
  1. Inspect mechanical key condition; check operation of ignition switch; check operation of indicator lights, warning lights and/or alarms; check instruments; record oil pressure and system voltage; check operation of electronic power take-off (PTO) and engine idle speed controls (if applicable). (IMMR, TST, MTST) P-1.
  2. Check operation of all accessories. (IMMR, TST, MTST) P-1.

- Workplace skills: Section I

Tools and Materials: A drivable medium- or heavy-duty truck; a yard not on public property; and a marked parking bay

Protective Clothing: Standard shop apparel, including coveralls or shop coat, safety glasses, and safety footwear

PROCEDURE

1. Identify the truck by recording the following vehicle data:

   OEM _____ Model year _____ Engine _____ VIN _____

   Task completed _____

2. Ensure that no persons or objects are under the truck before starting the engine.

   Task completed _____

3. Enter the vehicle cab. Adjust the seat to your height. Observe the location of the driver controls. Check the visibility through 360 degrees. Adjust the mirrors if necessary.

   Task completed _____

4. Start the engine. Build up the air pressure to cut-out. Release the parking brake. Put the truck into gear and move forward just inches. Apply the service brakes using the treadle valve. The truck should stop abruptly. Hold the service brakes in the applied position for a few moments, listening and observing (application gauge) for major air leaks.

    Task completed _____

5. Release the service brakes. Move the vehicle forward and drive to the designated parking area.

    Task completed _____

6. If the designated parking area is inside a shop, stop the truck outside the door. Quickly check that the door is fully open and that there is sufficient clearance to safely move the vehicle through the doorway. Also check that the designated bay is clear of people and objects.

    Task completed _____

7. Park the truck. Engage the parking brakes. Switch off the engine. For additional safety chock a pair of wheels on the vehicle.

    Task completed _____

8. Fit an exhaust extraction pipe to the exhaust stack(s). This is required to minimize the amount of exhaust fumes discharged into the closed shop when the engine is restarted.

    Task completed _____

9. When removing the truck from the bay, the preceding procedure can be reversed. However, the engine should be started and the air pressure built to system cut-out before removing the exhaust extraction stack(s).

    Task completed _____

STUDENT SELF-EVALUATION

Check	Level	Competency	Comments
	4	Mastered task	
	3	Competent but need further help	
	2	Needed a lot of help	
	0	Did not understand the task	

INSTRUCTOR EVALUATION

Check	Level	Competency	Comments
	4	Mastered task	
	3	Competent but needs further help	
	2	Requires more training	
	0	Unable to perform task	

ONLINE TASKS

Use a search engine to access the following Web pages and describe how each organization could be important to a technician working in the truck repair industry.

1. The American Trucking Association (ATA)
2. The Society of Automotive Engineers (SAE)
3. Occupational Safety and Health Administration (OSHA)
4. Technology and Maintenance Council (TMC)

STUDY TIPS

Identify 5 key points in Chapter 2. Try to be as brief as possible.

Key point 1 _____

Key point 2 _____

Key point 3 _____

Key point 4 _____

Key point 5 _____

3 Tools and Fasteners

Objectives

After reading this chapter, you should be able to:
- List some of the common hand tools used in heavy-duty truck repair.
- Describe how to use common pneumatic, electrical, and hydraulic power tools used in heavy-duty truck repair.
- Identify the mechanical and electronic measuring tools used in the heavy-duty truck shop.
- Describe the proper procedure for measuring with a micrometer.
- Read standard and metric micrometers.
- Use LOTO devices appropriately.
- Identify the types of manufacturer service literature used in truck repair facilities and describe the type of information each provides.
- Explain the principles and precautions of working with various heavy-duty truck fasteners.
- Identify SAE Grade 2, Grade 5, and Grade 8 fasteners.
- Outline the process for removing and installing rivets.
- Identify some common rivets.
- Describe the application of sealants and adhesives used in the trucking industry.

PRACTICE QUESTIONS

1. A wrench with open-end jaws at one end and a box-end jaw at the other, both of the same nominal dimension, is known as a(n):
   a. lineman wrench
   b. combination wrench
   c. spanner wrench
   d. adjustable wrench

2. Which of the following wrenches should be used to tighten pipe nuts that would be found on diesel fuel systems or hydraulic hoses?
   a. open-end wrench
   b. box-end wrench
   c. adjustable wrench
   d. line wrench

3. Which type of pliers would be most suitable for intricate electrical work?
   a. lineman pliers
   b. adjustable pliers
   c. channellock™ pliers
   d. needle nose pliers

4. Which of the following could also be called a *hex* socket?
    a. 4-point socket
    b. 6-point socket
    c. 8-point socket
    d. 12-point socket
5. Which of the following tools would be used to cut virgin threads on a steel shaft?
    a. tap
    b. die
    c. grommet
    d. thread file
6. How is an arbor press actuated?
    a. manually
    b. pneumatically
    c. hydraulically
    d. electrically
7. Through how many thimble rotations must a standard micrometer be turned to make the spindle move ¼ inch?
    a. 10
    b. 20
    c. 30
    d. 40
8. Through how many thimble rotations must a metric micrometer be turned to make the spindle move 12.5 mm?
    a. 1.25
    b. 12.5
    c. 25
    d. 50
9. Technician A says that it is good practice to replace all original Grade 5 fasteners with Grade 8 fasteners when performing suspension repairs. Technician B says that a Grade 8 rated Huck fastener may be replaced by a Grade 8 nut and bolt. Who is correct?
    a. Technician A only
    b. Technician B only
    c. both A and B
    d. neither A nor B
10. Which of the following media is capable of holding the most amount of information using the least amount of space?
    a. workshop manuals
    b. hard copy technical service bulletins (TSBs)
    c. CD
    d. USB flash drive

JOB SHEET 3.1

Name _____ Station _____ Date _____

Metric-to-Standard Conversion Familiarization.

Performance Objective(s): Become familiar with using metric and standard micrometers and converting the measured values from one system to the other.

ASE Education Foundation Correlation

This job sheet addresses the following ASE Education Foundation task(s):

- Workplace skills: Section K
- Narrative for science-related academic skills
- Narrative for mathematics-related academic skills

Tools and Materials: Metric and standard micrometers and an assortment of both metric and SAE capscrews

Protective Clothing: If this exercise is performed in a shop or garage setting, standard shop equipment should be worn. This should include safety boots/shoes, coveralls, and eye protection. The exercise may also be performed in a classroom.

PROCEDURE

1. Obtain four different SAE bolts sized between ¼ inch and 1 inch. Measure the hex caps on each using a standard outside micrometer. Determine what size open-jaw wrench is required for each capscrew.

   Task completed _____

2. Convert each SAE wrench size into a metric value and average it to the nearest 0.10 mm.

   Task completed _____

3. Obtain four different metric bolts sized between 6 mm and 13 mm. Measure the hex caps on each using a metric outside micrometer. Determine what size open-jaw wrench would be required for each capscrew.

   Task completed _____

4. Convert each metric wrench size into a standard value and average it to the nearest inch measurement.

   Task completed _____

STUDENT SELF-EVALUATION

Check	Level	Competency	Comments
	4	Mastered task	
	3	Competent but need further help	
	2	Needed a lot of help	
	0	Did not understand the task	

INSTRUCTOR EVALUATION

Check	Level	Competency	Comments
	4	Mastered task	
	3	Competent but needs further help	
	2	Requires more training	
	0	Unable to perform task	

JOB SHEET 3.2

Name _____ Station _____ Date _____

Truck Technician's Toolbox Inventory.

Performance Objective(s): Build a paper inventory of tools required by an entry-level truck technician using spreadsheet or word processing software. You can use a specified value—for instance, $3,000—to make this exercise more challenging. Another way of doing this is to make two lists using different total dollar values, for example, $2,000 versus $5,000. Note how your priorities change.

ASE Education Foundation Correlation

This job sheet addresses the following ASE Education Foundation task(s):

- Narrative for language arts-related studies

Tools and Materials: Tool supplier catalogs and a computer with either a word processing or spreadsheet program

Protective Clothing: None required

PROCEDURE

1. Obtain at least two tool supplier catalogs with price lists and the *Heavy-Duty Truck Systems, Sixth Edition,* textbook to use as a reference.

   Task completed _____

2. Select a spreadsheet or word processing program and build an inventory of "most needed" tools that can be obtained within the budget target.

   Task completed _____

3. Obtain a printout of the tool inventory.

   Task completed _____

4. Compare the contents of your tool inventory with those of other students in your class. Make corrections based on the discussions that result.

   Task completed _____

JOB SHEET 3.3

Name _____ Station _____ Date _____

Precision Measuring Tool Familiarization.

Performance Objective(s): Learn to recognize the components and make measurements utilizing commonly used measuring tools.

ASE Education Foundation Correlation

This job sheet addresses the following ASE Education Foundation task(s):

III. Brakes
 C. Air Brakes: Mechanical/Foundation Brake System
  4. Inspect rotor and mounting surface; measure rotor thickness, thickness variation, and lateral runout; determine needed action. (IMMR, TST, MTST) P-1.
  5. Inspect, clean, and adjust air disc brake caliper assemblies; inspect and measure disc brake pads; inspect mounting hardware; perform needed action. (IMMR, TST, MTST) P-1.
  6. Remove brake drum; clean and inspect brake drum and mounting surface; measure brake drum diameter; measure brake lining thickness; inspect brake lining condition; determine needed action. (IMMR, TST, MTST) P-1.

- Narrative for mathematics-related academic skills

Tools and Materials: Feeler gauges, Vernier calipers, outside micrometer, inside micrometer, depth micrometer, small hole gauge, telescoping gauges

Protective Clothing: None required

PROCEDURE

For each of the following figures, list three common uses.

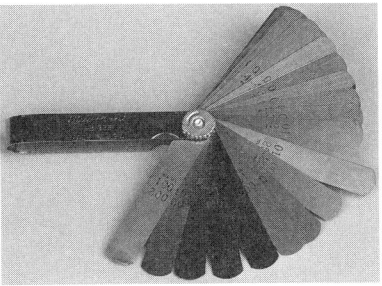

FIGURE 3–1 Thickness gauges

1. _____
2. _____
3. _____

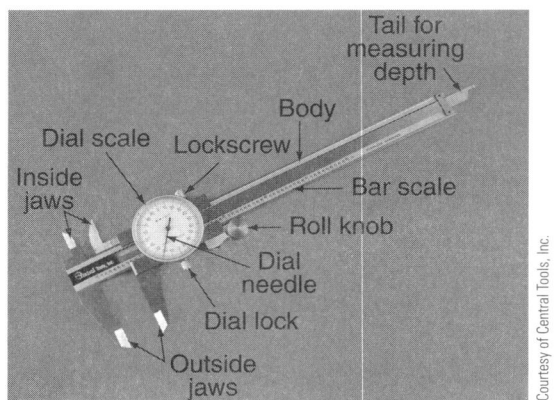

FIGURE 3–2 Vernier calipers with dial scale

1. _____
2. _____
3. _____

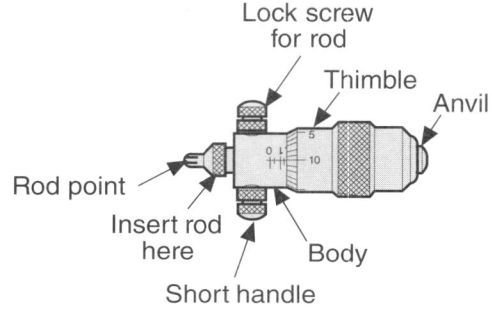

FIGURE 3–3 Inside micrometer terminology

1. _____
2. _____
3. _____

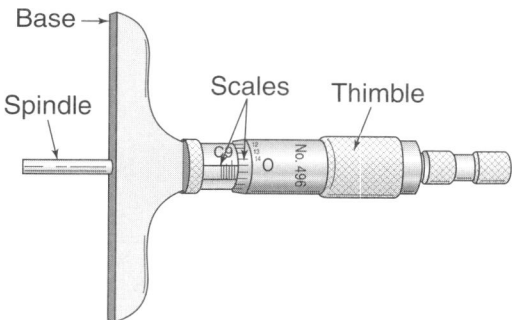

FIGURE 3–4 Depth micrometer terminology

1. _____
2. _____
3. _____

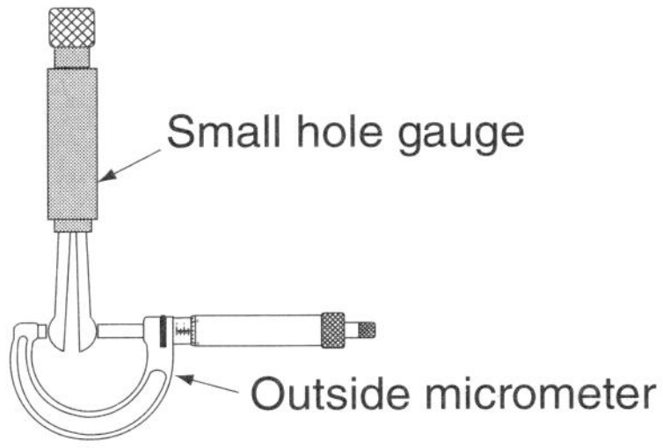

FIGURE 3–5 Small hole gauge

1. _____
2. _____
3. _____

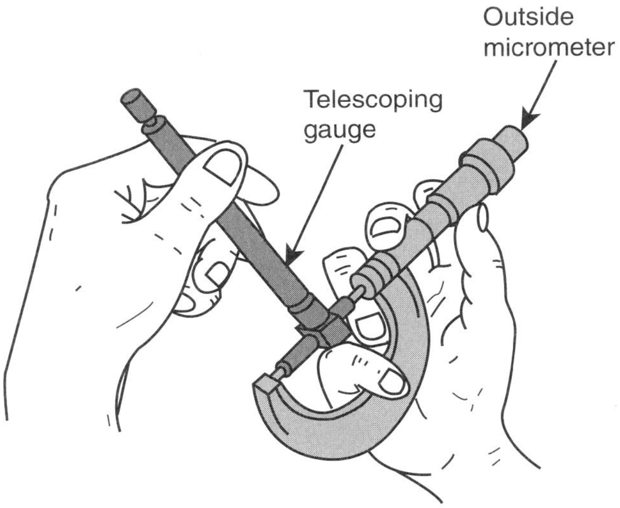

FIGURE 3–6 Telescoping gauge

1. _____
2. _____
3. _____

Tools and Fasteners 25

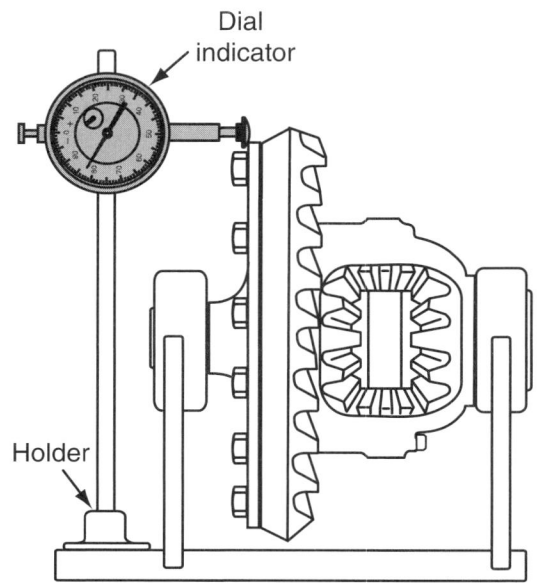

FIGURE 3-7 Dial indicator set to measure axial runout

Axial runout:

1. _____
2. _____
3. _____

Radial runout:

1. _____
2. _____
3. _____

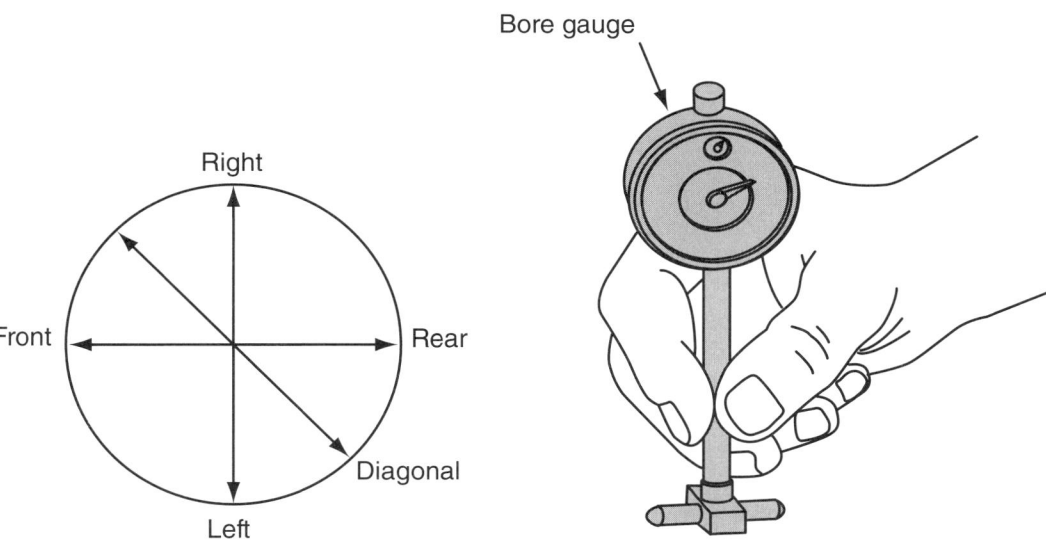

FIGURE 3-8 Dial bore gauge

1. _____
2. _____
3. _____

Task completed _____

STUDENT SELF-EVALUATION

Check	Level	Competency	Comments
	4	Mastered task	
	3	Competent but need further help	
	2	Needed a lot of help	
	0	Did not understand the task	

INSTRUCTOR EVALUATION

Check	Level	Competency	Comments
	4	Mastered task	
	3	Competent but needs further help	
	2	Requires more training	
	0	Unable to perform task	

JOB SHEET 3.4

Name _____ Station _____ Date _____

Reading a Standard Micrometer.

Performance Objective(s): Identify the components of a standard micrometer and interpret the readings.

ASE Education Foundation Correlation

This job sheet addresses the following ASE Education Foundation task(s):

- Narrative for mathematics-related academic skills

Tools and Materials: Outside micrometer

Protective Clothing: None required

PROCEDURE

1. Referencing Figure 3–9, identify the components of a standard outside micrometer by filling in the blanks. Try to do this without referencing the textbook.

   Task completed _____

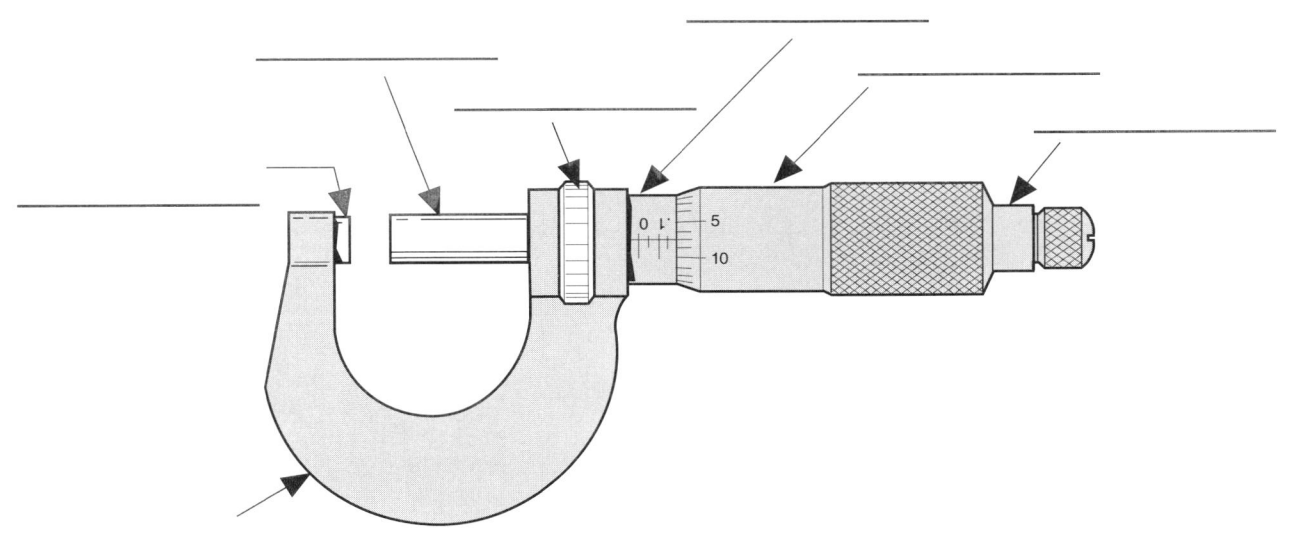

FIGURE 3–9 Nomenclature identification for a standard outside micrometer

Tools and Fasteners   29

2. Referencing Figure 3–10, fill in the readings below each figure.

Task completed _____

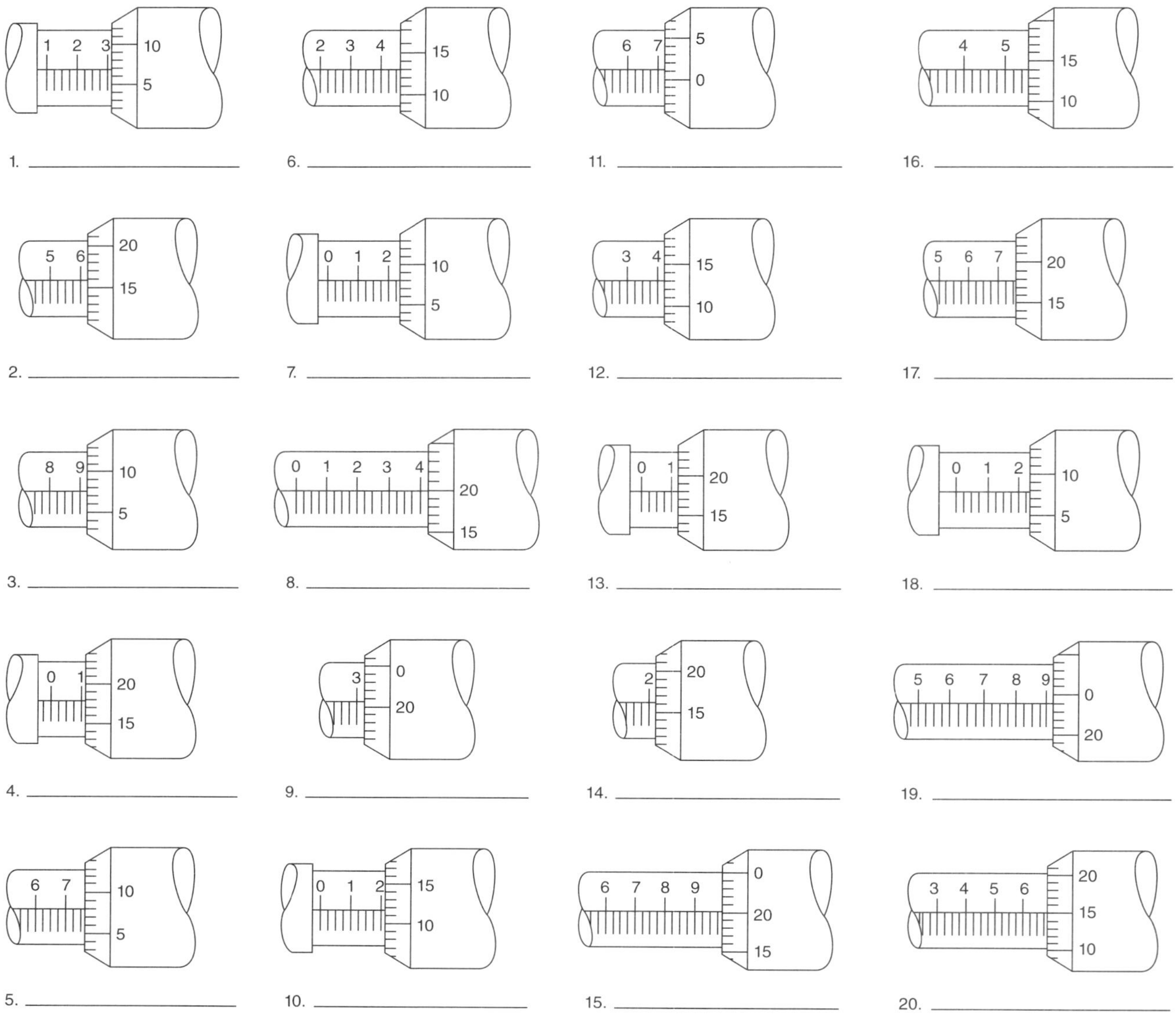

1. _____ 6. _____ 11. _____ 16. _____

2. _____ 7. _____ 12. _____ 17. _____

3. _____ 8. _____ 13. _____ 18. _____

4. _____ 9. _____ 14. _____ 19. _____

5. _____ 10. _____ 15. _____ 20. _____

FIGURE 3–10 Standard micrometer exercise

STUDENT SELF-EVALUATION

Check	Level	Competency	Comments
	4	Mastered task	
	3	Competent but need further help	
	2	Needed a lot of help	
	0	Did not understand the task	

INSTRUCTOR EVALUATION

Check	Level	Competency	Comments
	4	Mastered task	
	3	Competent but needs further help	
	2	Requires more training	
	0	Unable to perform task	

ONLINE TASKS

Use a search engine to locate and access the following Web pages and discuss how effectively each company presents itself online:

1. Ace Tools
2. Armstrong Tools
3. Blackhawk Tools
4. Craftsman Tools
5. Mac Tools
6. Snap-on Tools
7. Stanley Tools
8. Starrett Tools

Now check out the following American standards organizations online:

1. NPT
2. USS
3. SAE
4. ISO

STUDY TIPS

Identify 5 key points in Chapter 3. Try to be as brief as possible.

Key point 1 _____

Key point 2 _____

Key point 3 _____

Key point 4 _____

Key point 5 _____

4 Maintenance Programs

Objectives

After reading this chapter, you should be able to:

- Explain the characteristics and benefits of a well-planned maintenance program.
- Define the terms *preventive maintenance* (PM), *condition-based maintenance* (CBM), and *predictive-based maintenance* (PBM)
- Explain how trucking companies are increasingly using prognostics to reduce maintenance and service costs.
- Define VRMS and explain how it is used by the truck service industry.
- Interpret some basic VRMS codes relating to service procedures and components.
- List and describe the steps of the pretrip inspection procedure.
- Identify the different categories of preventive maintenance schedules.
- Describe the criteria for deadlining or out-of-service (OOS) tagging a vehicle.
- Implement a preventive maintenance schedule that conforms to federal inspection regulations.
- Outline inspector qualifications and record-keeping requirements.
- Select the correct lubricants for servicing trucks on preventive maintenance schedules.
- Describe the operation of on-board chassis lube systems.
- Prepare trucks and trailers for cold weather by winterizing.
- Describe how electronics have changed the modern maintenance environment.

PRACTICE QUESTIONS

1. Which of the following trailer chassis systems topped the list of maintenance concerns of fleet managers?
   a. swing doors
   b. suspensions
   c. tires
   d. brakes
2. On which of the following documents does the truck driver have input to preventive maintenance inspections?
   a. Schedule A report
   b. Schedule B report
   c. technician inspection report
   d. driver inspection report
3. Which of the following words is used in trucking to mean the same thing as out-of-service (OOS)?
   a. downtime
   b. deadlined
   c. beyond repair
   d. red-tagged

4. Which of the following would represent the OOS specification for a fifth wheel pivot pin and bracket?
   a. ¼ inch
   b. ⅜ inch
   c. ½ inch
   d. ¾ inch
5. Which of the following describes an OOS tire specification?
   a. 2/32 inch or less measured in any groove
   b. 2/32 inch or less measured anywhere in two adjacent grooves
   c. 2/16 inch or less measured in any groove
   d. 3/16 inch or more measured in any groove
6. Which type of diesel engine oil classification would conform to OEM service requirements for a four-stroke cycle engine equipped with a diesel particulate filter?
   a. CF-2
   b. CC
   c. CI-4
   d. CJ-4
7. Which of the following gear lube classifications would be appropriate for use in most OEM differential carrier assemblies?
   a. API GL-4
   b. API GL-5
   c. API GL-6
   d. API GL-7
8. Which of the following acronyms refers to types of antifreeze?
   a. EP
   b. EG
   c. PG
   d. PT
9. Which of the following devices would represent the heaviest load on a chassis electrical system when running?
   a. headlights
   b. heater fan
   c. electric wipers
   d. engine electronics
10. Which of the following terms correctly describes a standard grease fitting?
    a. Schrader valve
    b. Zerk fitting
    c. ball fitting
    d. pin valve

MAINTENANCE INSPECTION SHEETS

The lab task sheets in this section focus on maintenance inspection check sheets. Properly performing maintenance check sheets have a way of developing a truck technician's eye for detail that can be invaluable. Sound maintenance practices are key to avoiding costly, unplanned downtime in truck fleets. A maintenance checklist provides the root paper trail from which a breakdown can be either entirely avoided or blame-apportioned when running failures occur.

Most of our largest fleets provide specialty training in maintenance practices for their technicians. This is usually preventive maintenance (PM) based. In addition, most fleets are looking at condition-based maintenance (CBM) and the science of prognostics. Prognostics relate to practices of observing component performance through vehicle life span and replacing components only when necessary. This type of predictive maintenance can save unnecessary expenditure on PM.

It is important for entry-level technicians to remember that when checking off any field in a maintenance inspection checklist, they are affirming that they have properly inspected and/or made the required repair. Some of this work can be tedious but that is no excuse for skipping inspection areas. The effects of sloppy MAINTENANCE INSPECTION are costly to fleets, and in the competitive environment of the trucking industry; there is no room for fleet technicians who cannot properly perform maintenance inspections.

There follows a number of different types of truck inspection checklists, ranging from a driver's pre-/post-trip circle check to detailed C-category inspections. Many of the tasks are repeated but you should quickly get a feel for what constitutes each type of inspection. The checklists used here are sourced primarily from OEM recommended practices, but if you investigate you will discover that most transport fleets adapt these recommendations to make them specific for their own operating formats.

JOB SHEET 4.1

Name _____ Station _____ Date _____

Perform a Pretrip Circle Check on a Tractor/Trailer Combination.

Performance Objective(s): Use a typical circle check (pretrip inspection) to familiarize yourself with the routine inspection that a truck driver is required to make before each trip.

ASE Education Foundation Correlation

This job sheet addresses the following ASE Education Foundation task(s):

I. Diesel Engines
   A. General
   1. Research vehicle service information, including fluid type, vehicle service history, service precautions, and technical service bulletins. (IMMR, TST, MTST) P-1.
   4. Check engine operation (starting and running) including: noise, vibration, smoke, etc., determine needed action. (IMMR, TST) P-2.
   4. Diagnose engine operation (starting and running) including: noise, vibration, smoke, etc., determine needed action. (MTST) P-2.

   D. Lubrication Systems
   2. Check engine oil level, condition, and consumption; take engine oil sample; determine needed action. (IMMR, TST, MTST) P-1.

   E. Cooling System
   1. Check engine coolant type, level, condition, and test coolant for freeze protection and additive package concentration. (IMMR, TST, MTST) P-1.

   F. Air Induction and Exhaust Systems
   2. Check air induction system including: cooler assembly, piping, hoses, clamps, and mountings; replace air filter as needed; reset restriction indicator (if applicable). (IMMR, TST) P-1.
   2. Diagnose air induction system problems; inspect, clean, and/or replace cooler assembly, piping, hoses, clamps, and mountings; replace air filter as needed; reset restriction indicator (if applicable). (MTST) P-1.

   G. Fuel System
   1. Check fuel level and condition; determine needed action. (IMMR, TST, MTST) P-1.
   2. Inspect fuel tanks, vents, caps, mounts, valves, screens, crossover system, hoses, lines, and fittings; determine needed action. (IMMR, TST, MTST) P-1.

VII. Cab
   D. Hardware
   1. Check operation of wipers and washer; inspect windshield glass for cracks or discoloration; check sun visor; check seat condition, operation, and mounting; check door glass and window operation; verify operation of door and cab locks; inspect steps and grab handles; inspect mirrors, mountings, brackets, and glass. (IMMR) P-1.
   1. Test operation of wipers and washer; inspect windshield glass for cracks or discoloration; check sun visor; check seat condition, operation, and mounting; check door glass and window operation; verify operation of door and cab locks; inspect steps and grab handles; inspect mirrors, mountings, brackets, and glass; determine needed action. (TST, MTST) P-1.

Tools and Materials: A highway tractor coupled with a semi-trailer, standard personal and shop tools, the following circle check form, and a pen or pencil

Protective Clothing: Standard shop apparel including coveralls or shop coat, safety glasses, and safety footwear

PROCEDURE
Ensure that the shop LOTO is observed.

Complete a circle check inspection referencing Figure 4–1.

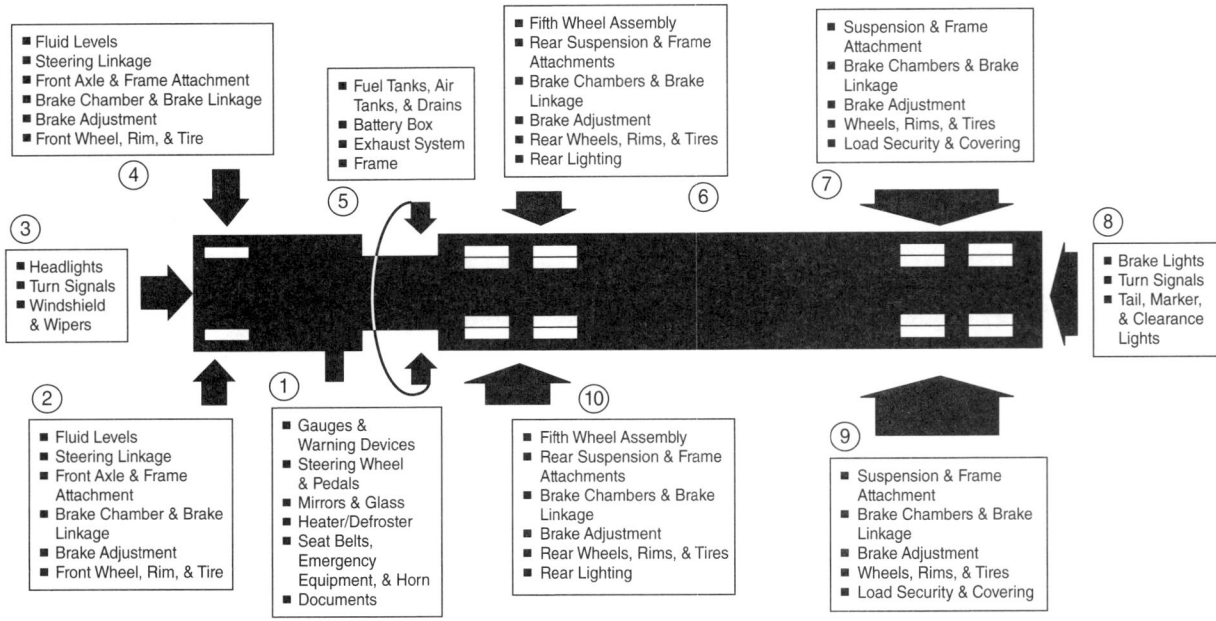

FIGURE 4–1 Circle check on a tractor/semi-trailer combination

STATION 1: CAB

Check	Ok	Corrected
Gauges/warning devices		
Steering wheel/pedals		
Mirrors and glass		
Heater/defroster		
Seat belts and horn		
Emergency equipment		
Documents		

STATION 2: LEFT FRONT TRACTOR

Check	Ok	Corrected
Fluid levels		
Steering linkage		
Front axle/front suspension		
Brake chamber/linkage		
Brake stroke		
Front wheel/rim/tire		

STATION 3: FRONT TRACTOR

Check	Ok	Corrected
Headlights		
Turn signals		
Windshield and wipers		

STATION 4: RIGHT FRONT TRACTOR

Check	Ok	Corrected
Fluid levels		
Air filter inlet restriction		
Front axle/front suspension		
Brake chamber/linkage		
Brake stroke		
Front wheel/rim/tire		

STATION 5: RIGHT TRACTOR

Check	Ok	Corrected
Fuel tanks, air tanks and drains		
Battery banks		
Exhaust system		
Frame		
Cross-members		

STATION 6: RIGHT REAR TRACTOR/RIGHT FRONT TRAILER

Check	Ok	Corrected
Fifth wheel/upper coupler		
Rear suspension/frame		
Brake chambers/linkages		
Brake stroke/adjustment		
Rear bogies/wheels/tires		
Rear lighting		

STATION 7: RIGHT SIDE TRAILER

Check	Ok	Corrected
Suspension and frame		
Brake chambers/linkage		
Brake adjustment		
Wheels/rims/tires		
Load security and covering		

STATION 8: REAR TRAILER

Check	Ok	Corrected
Brake lights		
Turn signals		
Tail/marker/clearance lights		
Mudflaps		

STATION 9: LEFT SIDE TRAILER

Check	Ok	Corrected
Suspension and frame		
Brake chambers/linkage		
Brake adjustment		
Wheels/rims/tires		
Load security and covering		

STATION 10: LEFT REAR TRACTOR/LEFT FRONT TRAILER

Check	Ok	Corrected
Fifth wheel/upper coupler		
Rear suspension/frame		
Brake chambers/linkages		
Brake stroke/adjustment		
Rear bogies/wheels/tires		
Rear lighting		

STUDENT SELF-EVALUATION

Check	Level	Competency	Comments
	4	Mastered task	
	3	Competent but need further help	
	2	Needed a lot of help	
	0	Did not understand the task	

INSTRUCTOR EVALUATION

Check	Level	Competency	Comments
	4	Mastered task	
	3	Competent but need further help	
	2	Requires more training	
	0	Unable to perform task	

JOB SHEET 4.2

Name _____ Station _____ Date _____

Perform a Generic Maintenance Inspection on a Linehaul Truck.

Performance Objective(s): Use the checklist that follows to perform a generic PM inspection. The objective is to get a general idea of what is required in A/B/C inspections. In a later exercise, we will take a look at an OEM approach to performing a similar maintenance inspection.

ASE Education Foundation Correlation

This job sheet addresses the following ASE Education Foundation task(s):

I. **Diesel Engines**
   A. General
   1. Research vehicle service information, including fluid type, vehicle service history, service precautions, and technical service bulletins. (IMMR, TST, MTST) P-1.
   4. Check engine operation (starting and running) including: noise, vibration, smoke, etc., determine needed action. (IMMR, TST) P-2.
   4. Diagnose engine operation (starting and running) including: noise, vibration, smoke, etc., determine needed action. (MTST) P-2.

   D. Lubrication Systems
   2. Check engine oil level, condition, and consumption; take engine oil sample; determine needed action. (IMMR, TST, MTST) P-1.

   E. Cooling System
   1. Check engine coolant type, level, condition, and test coolant for freeze protection and additive package concentration. (IMMR, TST, MTST) P-1.

   F. Air Induction and Exhaust Systems
   2. Check air induction system including: cooler assembly, piping, hoses, clamps, and mountings; replace air filter as needed; reset restriction indicator (if applicable). (IMMR, TST) P-1.
   2. Diagnose air induction system problems; inspect, clean, and/or replace cooler assembly, piping, hoses, clamps, and mountings; replace air filter as needed; reset restriction indicator (if applicable). (MTST) P-1.

   G. Fuel System
   1. Check fuel level and condition; determine needed action. (IMMR, TST, MTST) P-1.
   2. Inspect fuel tanks, vents, caps, mounts, valves, screens, crossover system, hoses, lines, and fittings; determine needed action. (IMMR, TST, MTST) P-1.
   4. Replace fuel filter; prime and bleed fuel system. (IMMR, TST, MTST) P-1.

II. **Drive Train**
   A. General
   2. Identify drive train components, transmission type, and configuration. (IMMR, TST, MTST) P-1.

   B. Clutch
   1. Inspect and adjust clutch, clutch brake, linkage, cables, levers, brackets, bushings, pivots, springs, and clutch safety switch (includes push-type and pull-type); check pedal height and travel; determine needed action. (IMMR, TST, MTST) P-1.
   2. Inspect clutch master cylinder fluid level; check clutch master cylinder, slave cylinder, lines, and hoses for leaks and damage; determine needed action. (IMMR, TST, MTST) P-1.

   C. Transmission
   4. Check transmission fluid level and condition; determine needed action. (IMMR, TST, MTST) P-1.
   5. Inspect transmission breather; inspect transmission oil filters, coolers and related components; determine needed action. (IMMR, TST, MTST) P-2.

D. Driveshaft and Universal Joints
1. Inspect, service, and/or replace driveshafts, slip joints, yokes, drive flanges, support bearings, universal joints, boots, seals, and retaining/mounting hardware; check phasing of all shafts. (IMMR, TST, MTST) P-1.

E. Drive Axles
1. Check for fluid leaks; inspect drive-axle housing assembly, cover plates, gaskets, seals, vent/breather, and magnetic plugs. (IMMR) P-1.
1. Check and repair fluid leaks; inspect drive-axle housing assembly, cover plates, gaskets, seals, vent/breather, and magnetic plugs. (TST, MTST) P-1.
2. Check drive-axle fluid level and condition; check drive-axle filter; determine needed action. (IMMR, TST, MTST) P-1.

III. Brakes
A. General
1. Research vehicle service information, including fluid type, vehicle service history, service precautions, and technical service bulletins. (IMMR, TST, MTST) P-1.
2. Identify brake system components and configurations (including air and hydraulic systems, parking brake, power assist, and vehicle dynamic brake systems). (IMMR, TST, MTST) P-1.

B. Air Brakes: Air Supply and Service Systems
1. Inspect air supply system components such as compressor, governor, air drier, tanks, and lines; inspect service system components such as lines, fittings, mountings, and valves (hand brake/trailer control, brake relay, quick release, tractor protection, emergency/spring brake control/modulator, pressure relief/safety); determine needed action. (IMMR, TST, MTST) P-1.
2. Verify proper gauge operation and readings; verify low pressure warning alarm operation; perform air supply system tests such as pressure build-up, governor settings, and leakage; drain air tanks and check for contamination. (IMMR) P-1.
2. Test gauge operation and readings; test low pressure warning alarm operation; perform air supply system tests such as pressure build-up, governor settings, and leakage; drain air tanks and check for contamination; determine needed action. (TST, MTST) P-1.

C. Air Brakes: Mechanical/Foundation Brake Systems
1. Inspect service brake chambers, diaphragms, clamps, springs, pushrods, clevises, and mounting brackets; determine needed action. (IMMR) P-1.
1. Inspect and test service brake chambers, diaphragms, clamps, springs, pushrods, clevises, and mounting brackets; determine needed action. (TST) P-1.
1. Inspect, test, repair, and/or replace service brake chambers, diaphragms, clamps, springs, pushrods, clevises, and mounting brackets; determine needed action. (MTST) P-1.
2. Identify slack adjuster type; inspect slack adjusters; determine needed action. (IMMR) P-1.
2. Identify slack adjuster type; inspect slack adjusters; perform needed action. (TST, MTST) P-1.
3. Check camshafts (S-cams), tubes, rollers, bushings, seals, spacers, retainers, brake spiders, shields, anchor pins, and springs; determine needed action. (IMMR) P-1.
3. Check camshafts (S-cams), tubes, rollers, bushings, seals, spacers, retainers, brake spiders, shields, anchor pins, and springs; perform needed action. (TST, MTST) P-1.
4. Inspect rotor and mounting surface; measure rotor thickness, thickness variation, and lateral runout; determine needed action. (IMMR, TST, MTST) P-1.
5. Inspect, clean, and adjust air disc brake caliper assemblies; inspect and measure disc brake pads; inspect mounting hardware; perform needed action. (IMMR, TST, MTST) P-1.
6. Remove brake drum; clean and inspect brake drum and mounting surface; measure brake drum diameter; measure brake lining thickness; inspect brake lining condition; determine needed action. (IMMR, TST, MTST) P-1.

D. Air Brakes: Parking Brake System
1. Inspect and check parking (spring) brake chamber for leaks; determine needed action. (IMMR) P-1.
1. Inspect, test, and/or replace parking (spring) brake chamber. (TST, MTST) P-1.

E. Hydraulic Brakes: Hydraulic System
1. Check master cylinder fluid level and condition; determine proper fluid type for application. (IMMR, TST, MTST) P-1.

2. Inspect hydraulic brake system components for leaks and damage. (IMMR) P-1.
2. Inspect hydraulic brake system for leaks and damage; test, repair, and/or replace hydraulic brake system components. (TST, MTST) P-1.
3. Check hydraulic brake system operation including pedal travel, pedal effort, and pedal feel; determine needed action. (IMMR, TST, MTST) P-1.

G. Hydraulic Brakes: Parking Brake System
1. Check parking brake operation; inspect parking brake application and holding devices. (IMMR) P-1.
1. Check parking brake operation; inspect parking brake application and holding devices; adjust, repair, and/or replace as needed. (TST, MTST)

I. Vehicle Dynamic Brake Systems (Air and Hydraulic): Antilock Brake System (ABS), Automatic Traction Control (ATC) System, and Electronic Stability Control (ESC) System
1. Observe antilock brake system (ABS) warning light operation including trailer and dash mounted trailer ABS warning light; determine needed action. (IMMR, TST, MTST) P-1.
2. Observe automatic traction control (ATC) and electronic stability control (ESC) warning light operation; determine needed action. (IMMR, TST, MTST) P-1.

J. Wheel Bearings
1. Clean, inspect, lubricate, and/or replace wheel bearings and races/cups; replace seals and wear rings; inspect spindle/tube; inspect and replace retaining hardware; adjust wheel bearings; check hub assembly fluid level and condition; verify end play with dial indicator method. (IMMR, TST, MTST) P-1.
2. Identify, inspect, and/or replace unitized/preset hub bearing assemblies. (IMMR, TST, MTST) P-2.

IV. Suspension and Steering

B. Steering Column
1. Check steering wheel for free play, binding, and proper centering; inspect and service steering shaft U-joint(s), slip joint(s), bearings, bushings, and seals; phase steering shaft. (IMMR, TST, MTST) P-1.

C. Steering Pump and Gear Units
1. Check power steering pump and gear operation, mountings, lines, and hoses; check fluid level and condition; service filter; inspect system for leaks. (IMMR, TST, MTST) P-1.
2. Flush and refill power steering system; purge air from system. (IMMR, TST, MTST) P-2.

D. Steering Linkage
1. Inspect tie rod ends, ball joints, kingpins, pitman arms, idler arms, and other steering linkage components; lubricate as needed. (IMMR) P-1.
1. Inspect, service, repair, and/or replace tie rod ends, ball joints, kingpins, pitman arms, idler arms, and other steering linkage components. (TST, MTST) P-1.

E. Suspension Systems
1. Inspect shock absorbers, bushings, brackets, and mounts; determine needed action. (IMMR) P-1.
1. Inspect, service, repair, and/or replace shock absorbers, bushings, brackets, and mounts. (TST, MTST) P-1.
2. Inspect leaf springs, center bolts, clips, pins, bushings, shackles, U-bolts, insulators, brackets, and mounts; determine needed action. (IMMR) P-1.
2. Inspect, repair, and/or replace leaf springs, center bolts, clips, pins, bushings, shackles, U-bolts, insulators, brackets, and mounts. (TST, MTST) P-1.
5. Inspect and test air suspension pressure regulator and height control valves, lines, hoses, dump valves, and fittings; check and record ride height. (IMMR) P-1.
6. Inspect, test, repair, and/or replace air suspension pressure regulator and height control valves, lines, hoses, dump valves, and fittings; check and record ride height. (TST, MTST) P-1.

F. Wheel Alignment Diagnosis and Repair
3. Check and record camber. (TST, MTST) P-2.
4. Check and record caster. (TST, MTST) P-2.
5. Check, record, and adjust toe settings. (TST, MTST) P-1.
6. Check rear axle(s) alignment (thrustline/centerline) and tracking. (TST, MTST) P-2.
7. Identify turning/Ackerman angle (toe-out-on-turns) problems. (TST, MTST) P-3.
8. Check front axle alignment (centerline). (TST, MTST) P-2.

G. Wheels and Tires
1. Inspect tire condition; identify tire wear patterns; measure tread depth; verify tire matching (diameter and tread); inspect valve stem and cap; set tire pressure. (IMMR) P-1.
1. Inspect tire condition; identify tire wear patterns; measure tread depth; verify tire matching (diameter and tread); inspect valve stem and cap; set tire pressure; determine needed action. (TST, MTST) P-1.
3. Check wheel mounting hardware; check wheel condition; remove and install wheel/tire assemblies (steering and drive axle); torque fasteners to manufacturer's specification using torque wrench. (IMMR, TST, MTST) P-1.

H. Frame and Coupling Devices
1. Inspect, service, and/or adjust fifth wheel, pivot pins, bushings, locking mechanisms, mounting hardware, air lines, and fittings. (IMMR, TST, MTST) P-1.

V. Electrical/Electronic Systems
A. General
3. Demonstrate proper use of test equipment when measuring source voltage, voltage drop (including grounds), current flow, continuity, and resistance. (IMMR, TST, MTST) P-1.
5. Use wiring diagrams to trace electrical/electronic circuits. (IMMR) P-1.
5. Use wiring diagrams during the diagnosis (troubleshooting) of electrical/electronic circuit problems. (TST, MTST) P-1.
9. Use appropriate electronic service tool(s) and procedures to check, record, and clear diagnostic codes; interpret digital multimeter (DMM) readings. (IMMR) P-2.
9. Use appropriate electronic service tool(s) and procedures to diagnose problems; check, record, and clear diagnostic codes; interpret digital multimeter (DMM) readings. (TST, MTST) P-2.

B. Battery System
1. Identify battery type and system configuration. (IMMR, TST, MTST) P-1.
2. Confirm proper battery capacity for application; perform battery state-of-charge test; perform battery capacity test, determine needed action. (IMMR, TST, MTST) P-1.
3. Inspect battery, battery cables, connectors, battery boxes, mounts, and hold-downs; determine needed action. (IMMR, TST, MTST) P-1.
4. Charge battery using appropriate method for battery type. (IMMR, TST, MTST) P-1.

C. Starting System
2. Perform starter circuit cranking voltage and voltage drop tests. (IMMR)
2. Perform starter circuit cranking voltage and voltage drop tests; determine needed action. (TST, MTST)
3. Inspect starter control circuit switches, relays, connectors, terminals, wires, and harnesses (including over-crank protection). (IMMR) P-1.
3. Inspect and test starter control circuit switches (key switch, push button, and/or magnetic switch), relays, connectors, terminals, wires, and harnesses (including over-crank protection); determine needed action. TST, MTST) P-1.
5. Perform starter current draw tests; determine needed action. (TST, MTST) P-3.

D. Charging System
5. Perform charging system voltage and amperage output tests; perform AC ripple test. (IMMR) P-1.
5. Perform charging system voltage and amperage output tests; perform AC ripple test; determine needed action. (TST, MTST) P-1.
6. Perform charging circuit voltage drop tests; determine needed action. (TST, MTST) P-1.

E. Lighting Systems
2. Test, replace, and aim headlights. (IMMR, TST, MTST) P-1.

F. Instrument Cluster and Driver Information Systems
1. Check gauge and warning indicator operation. (IMMR, TST, MTST) P-1.

VI. Heating, Ventilation, and Air Conditioning (HVAC)
A. General
2. Identify heating, ventilation, and air-conditioning (HVAC) components and configuration. (IMMR, TST, MTST) P-1.
6. Perform A/C system performance test; determine needed action. (MTST) P-1.

VII. Cab
B. Instruments and Controls
2. Check operation of all accessories. (IMMR, TST, MTST) P-1.
D. Hardware
1. Check operation of wipers and washer; inspect windshield glass for cracks or discoloration; check sun visor; check seat condition, operation, and mounting; check door glass and window operation; verify operation of door and cab locks; inspect steps and grab handles; inspect mirrors, mountings, brackets, and glass; determine needed action. (IMMR, TST, MTST) P-1.

Tools and Materials: A highway truck or tractor unit in running condition, standard personal and shop tools, the following maintenance procedure, and a pen or pencil

Protective Clothing: Standard shop apparel including coveralls or shop coat, safety glasses, and safety footwear

PROCEDURE
Ensure that the shop LOTO is observed.

1. Complete the following form, noting whether each task has actually been performed or reviewed.

Task completed _____

PMI RECORD SHEET (LIGHT-/MEDIUM-DUTY VEHICLES)
Prepare Separate Report for Each Vehicle Inspected

Technology & Maintenance Council (TMC)
950 N. Glebe Road, Arlington, VA 22203
To reorder: Call (866) 821-3468
http://www.atabusinesssolutions.com
© 2015—TMC/ATA   Item #: T0472

ATA BUSINESS SOLUTIONS

COMPANY NAME		DATE	MILEAGE	ENGINE HOURS
STREET ADDRESS		VIN		
CITY, STATE & ZIP		VEHICLE MAKE & MODEL		
INSPECTOR(S) NAME (Please Print)		INSPECTOR(S) SIGNATURE		
SUPERVISOR'S NAME (Please Print)		SUPERVISOR'S SIGNATURE		
DRIVER'S VEHICLE CONDITION REPORT INCLUDED	YES NO	DRIVER'S NAME (Please Print)		
REPAIR ORDER NO.		PMI LEVEL (Circle One) A B C Other:		

PMI RECORDING SYMBOLOGY: ✓=OK   D=Defect Found   N=Not Applicable   X=Repair made

In-cab Inspection

ITEM		A	B	C	Initial
101	Check ignition key and switch.			■	
102	Check warning lights and alarms.				
103	Check controls, switches and interior lights.				
104	Check permits and interior damage. Review DVIR, maint. reports, etc.				
105	Inspect seats, seat belts and sleeper restraints.				
106	Check cab electrical / non-electrical accessories and "add-on" devices.				
107	Inspect sleeper berth area lighting, vents, privacy curtain, and additional accessories (if applicable).		■		
108	Check horns.				
109	Check windshield, glass and cover accessories.				
110	Check wiper and washer operation.				
111	Check safety equipment and decals.				
112	Check clutch free travel (if applicable) and pedal pads.				

Test Drive Inspection

ITEM		A	B	C	Initial
201	Check starter operation. Record idle RPM_____ PTO RPM_____				
202	Check HVAC in cab / sleeper berth and controls.				
203	Check clutch brake operation (if applicable).			■	
204	Check steering wheel play for binding.				
205	Check automatic and automated manual transmission selector positions (if applicable).				
206	Check transmission shifting.		■		
207	Check engine brake, fan clutch and cruise control operation.				
208	Check steering wheel pull.				
209	Check vehicle speed governor.				
210	Observe exhaust for excessive smoke.				
211	Test foot brake operation.				

Air-Hydraulic Brake Inspection

ITEM		A	B	C	Initial
301	Check parking and trailer brake operations (if applicable).				
302	Test air leakage rate / static check (if applicable).				
303	Test air or hydraulic brake systems for leaks.				
304	Check hydraulic booster and master cylinder functions (if applicable).				
305	Test low air pressure warning alarm and signal (if applicable).		■		
306	Test parking air brake to operate automatically (if applicable).				
307	Check air governor and dryer drain valve operation (if applicable).				
308	Drain air tanks and inspect air tank assembly (if applicable).	■			

Walk-around and Open Hood Inspection (cont)

ITEM		A	B	C	Initial
401	Secure vehicle properly.				
402	Check and lube driver-side seat mechanism, door latches and key locks.	■			
403	Inspect driver-side cab and chassis componentry, including fuel tank assembly, door mirrors, and steps.				
404	Inspect exterior body for damage.				
405	Inspect driver-side steer tire inflation pressure and rim.				
406	Inspect driver-side steer-axle hub and kingpin.				
407	Inspect driver-side steer tire tread and sidewall.				
408	Check driver-side steer-axle suspension.				
409	Inspect hood assembly including grille / bug screen, spot mirrors, and headlight lenses.				
410	Inspect front bumper assembly including license plate and holder.				
411	Check all exterior lighting, headlight alignment, reflectors and reflective tape.				
412	Check and lube cabover and lift assembly (cabover vehicle only).			■	
413	Check steering wheel column and gear box assembly.				
414	Check any installed hydraulic fluid systems, including power steering, clutch assemblies (if applicable).				
415	Inspect the condition of all electrical harnesses, connections, lines and hoses, fittings, mountings and bolts for chaffing and fightness.				
416	Check engine mounts and fan hub.				
417	Inspect engine front end, air intake and cooling systems.				
418	Inspect engine belts, pulleys, front main seal, coolant pump and alternator assembly.				
419	Check coolant/filer and reservoir tank. Record freeze point_____ °F				
420	Pressure test cooling system and inspect for leaks.	■			
421	Inspect A/C system and components.				
422	Check automatic transmission of level (if applicable).				
423	Check and fill windshield washer fluid reservoir.			■	
424	Check engine air filter and assembly.				
425	Clean or replace HVAC filter(s).	■	■		
426	Inspect passenger-side steer tire inflation pressure and rim.				
427	Inspect passenger-side steer-axle hub and kingpin.				
428	Inspect passenger-side steer tire tread and sidewall.				
429	Check passenger-side steer-axle suspension.				
430	Check and lube passenger-side seat mechanism, door latches and key locks.	■			

FIGURE 4–2 An example of a PM form

Walk-around and Open Hood Inspection (cont)

ITEM		Inspection Level A	B	C	Initial
431	Inspect passenger-side cab and chassis componentry including fuel tank assembly, door mirrors, and steps.				
432	Inspect deck plate, cab mounts and suspension.				
433	Inspect passenger-side drive tires air pressures and rims.				
434	Inspect passenger-side drive tires head and sidewalls.				
435	Inspect mudflaps, brackets, and rear-mounted equipment.				
436	Inspect freight cargo-box doors and interior (if applicable).				
437	Inspect freight cargo-box auxiliary equipment (if applicable).			■	
438	Inspect fifth wheel, mounting, lines and locks (if applicable).				
439	Inspect driver-side drive tires inflation pressures and rims.				
440	Inspect driver-side drive tires tread and sidewalls.				
441	Inspect tractor-trailer connections (if applicable).				
442	Inspect and test battery electrical system.			■	
443	Inspect battery cables and electrical components.				

ELECTRICAL TEST RESULTS

Battery Pack Test — Open Voltage_____ V   Voltage Drop_____ V

Crank Voltage_____ V   Alternator Regulated Voltage_____ V

Crank Amperage_____ A   NOTES:

Exhaust and Emissions

ITEM		A	B	C	Initial
601	Inspect engine exhaust manifold and turbocharger assembly if applicable.			■	
602	Inspect closer valve and EGR system (it applicable).				
603	Inspect entire DEF system.				
604	Inspect entire exhaust system.				
605	Inspect emission systems including DPF, DOC, and SCR.			■	

Final Inspection

ITEM		A	B	C	Initial
701	Refill engine crankcase oil.	■			
702	Start engine and inspect for leaks.				
703	ECM / ECU Diagnostics Check max veh spd. max cruise spd.				
704	Recheck engine oil.			■	
705	Install updated PM sticker (if vehicle meets DOT regulations).	■			
706	Validate emission label (if applicable).			■	

PMI "C" Inspection Supplemental Items

ITEM		A	B	C	Initial
801	Change power steering fluid and filter.			■	
802	Change differential(s) oil.			■	
803	Change transmission filter and fluid.			■	
804	Draw transmission fluid sample.			■	
805	Draw differential oil sample(s).			■	
806	Remove and inspect steer-axle wheel bearings.			■	
807	Remove and inspect drive-axle wheel bearings.			■	
808	Adjust engine head valves.			■	
809	Adjust engine fuel injectors.			■	
810	Check steer- and drive-axle alignment.			■	
811	Service desiccant pack in air dryer.			■	
812	Flush and fill coolant system.			■	
813	Recover / Vacuum / Charge A/C System.			■	
814	Dram and fitter fuel tank(s).			■	
815	Change engine oil centrifuge filter.			■	
816	Change engine breather filter.			■	

Engine and Chassis Inspection

ITEM		Inspection Level A	B	C	Initial
501	Take engine oil sample and inspect fill tube/dipstick.				
502	Grease all fittings accessible from above the unit and grease fifth wheel surface (if applicable).				
503	Change fuel filters and drain fuel/water separator.	■		■	
504	Inspect engine fuel pump and fuel system assembly.				
505	Inspect air compressor and oil pump including components.				
506	Drain engine oil and change filters.			■	
507	Inspect crankcase breather for condition, mounting and leakage.				
508	Inspect chassis steering linkage assembly.				
509	Inspect steer-axle brakes and wheel seals.				
510	Grease all Zerk fittings accessible from under the unit.				
511	Inspect clutch assembly and check for proper adjustment (if applicable).	■		■	
512	Inspect transmission assembly and gear selector linkage (if applicable).				
513	Inspect driveshaft assembly.				
514	Inspect differential assembly(ies) and check oil level(s).	■			
515	Inspect drive-axle brakes and wheel seals.				
516	Inspect brake chambers and check applied brake stroke measurement (if applicable).				
517	Inspect rear chassis suspension assembly.				
518	Inspect frame rails and crossmembers.				

Fleet Specific PMI Items.

				Initials

FIGURE 4–2 (continued)

Follow-up Items and Repairs			
ITEM	CONDITION	Date	Initials

COMMENTS

_____
_____
_____
_____
_____
_____
_____
_____
_____
_____

RECORD TREAD DEPTH, TIRE INFLATION PRESSURE, BRAKE LINING / PAD THICKNESS & APPLIED BRAKE STROKE MEASUREMENT

D_____/32nds
P_____PSI / KPA
B_____/32nds
S_____in/cm

SL

D = Tread Depth
P = Tire Inflation Pressure
B = Brake Lining/Pad Thickness Prior to Adjustment
S = Applied Brake Stroke Measurement

SR

D_____/32nds
P_____PSI / KPA
B_____/32nds
S_____in/cm

D_____/32nds
P_____PSI / KPA

D1LO **D1L1**

D_____/32nds
P_____PSI / KPA
B_____/32nds
S_____in/cm

D_____/32nds
P_____PSI / KPA
B_____/32nds
S_____in/cm

D1R1 **D1RO**

D_____/32nds
P_____PSI / KPA

D_____/32nds
P_____PSI / KPA

D2LO **D2L1**

D_____/32nds
P_____PSI / KPA
B_____/32nds
S_____in/cm

D_____/32nds
P_____PSI / KPA
B_____/32nds
S_____in/cm

D2R1 **D2RO**

D_____/32nds
P_____PSI / KPA

S=Steer D=Drive
L=Left R=Right
O=Outside I=Inside

1=1st Drive Axle
2=2nd Drive Axle
(From VMRS Code Key 23)

FIGURE 4–2 (continued)

STUDENT SELF-EVALUATION

Check	Level	Competency	Comments
	4	Mastered task	
	3	Competent but need further help	
	2	Needed a lot of help	
	0	Did not understand the task	

INSTRUCTOR EVALUATION

Check	Level	Competency	Comments
	4	Mastered task	
	3	Competent but needs further help	
	2	Requires more training	
	0	Unable to perform task	

JOB SHEET 4.3

Name _____ Station _____ Date _____

Perform a Chassis Lube Inspection on a Linehaul Truck.

Performance Objective(s): Use the checklist that follows to perform a chassis lube on a linehaul truck. The objective is to familiarize yourself with this simple procedure that is performed thousands of times daily in fleet truck shops and dealerships.

ASE Education Foundation Correlation

This job sheet addresses the following ASE Education Foundation task(s):

I. Diesel Engines
A. General
1. Research vehicle service information, including fluid type, vehicle service history, service precautions, and technical service bulletins. (IMMR, TST, MTST) P-1.

D. Lubrication Systems
2. Check engine oil level, condition, and consumption; take engine oil sample; determine needed action. (IMMR, TST, MTST) P-1.

E. Cooling System
1. Check engine coolant type, level, condition, and test coolant for freeze protection and additive package concentration. (IMMR, TST, MTST) P-1.

F. Air Induction and Exhaust Systems
2. Check air induction system including: cooler assembly, piping, hoses, clamps, and mountings; replace air filter as needed; reset restriction indicator (if applicable). (IMMR, TST) P-1.
2. Diagnose air induction system problems; inspect, clean, and/or replace cooler assembly, piping, hoses, clamps, and mountings; replace air filter as needed; reset restriction indicator (if applicable). (MTST) P-1.

G. Fuel System
2. Inspect fuel tanks, vents, caps, mounts, valves, screens, crossover system, hoses, lines, and fittings; determine needed action. (IMMR, TST, MTST) P-1.
4. Replace fuel filter; prime and bleed fuel system. (IMMR, TST, MTST) P-1.

II. Drive Train
B. Clutch
2. Inspect clutch master cylinder fluid level; check clutch master cylinder, slave cylinder, lines, and hoses for leaks and damage; determine needed action. (IMMR, TST, MTST) P-1.

C. Transmission
4. Check transmission fluid level and condition; determine needed action. (IMMR, TST, MTST) P-1.
5. Inspect transmission breather; inspect transmission oil filters, coolers and related components; determine needed action. (IMMR, TST, MTST) P-2.

E. Drive Axles
1. Check for fluid leaks; inspect drive-axle housing assembly, cover plates, gaskets, seals, vent/breather, and magnetic plugs. (IMMR) P-1.
1. Check and repair fluid leaks; inspect drive-axle housing assembly, cover plates, gaskets, seals, vent/breather, and magnetic plugs. (TST, MTST) P-1.
2. Check drive-axle fluid level and condition; check drive-axle filter; determine needed action. (IMMR, TST, MTST) P-1.

III. Brakes
A. General
2. Identify brake system components and configurations (including air and hydraulic systems, parking brake, power assist, and vehicle dynamic brake systems). (IMMR, TST, MTST) P-1.

D. Air Brakes: Parking Brake System
1. Inspect and check parking (spring) brake chamber for leaks; determine needed action. (IMMR) P-1.
1. Inspect, test, and/or replace parking (spring) brake chamber. (TST, MTST) P-1.

E. Hydraulic Brakes: Hydraulic System
1. Check master cylinder fluid level and condition; determine proper fluid type for application. (IMMR, TST, MTST) P-1.
2. Inspect hydraulic brake system components for leaks and damage. (IMMR) P-1.
2. Inspect hydraulic brake system for leaks and damage; test, repair, and/or replace hydraulic brake system components. (TST, MTST) P-1.

J. Wheel Bearings
1. Clean, inspect, lubricate, and/or replace wheel bearings and races/cups; replace seals and wear rings; inspect spindle/tube; inspect and replace retaining hardware; adjust wheel bearings; check hub assembly fluid level and condition; verify end play with dial indicator method. (IMMR, TST, MTST) P-1.

IV. Suspension and Steering
C. Steering Pump and Gear Units
1. Check power steering pump and gear operation, mountings, lines, and hoses; check fluid level and condition; service filter; inspect system for leaks. (IMMR, TST, MTST) P-1.

G. Wheels and Tires
1. Inspect tire condition; identify tire wear patterns; measure tread depth; verify tire matching (diameter and tread); inspect valve stem and cap; set tire pressure. (IMMR) P-1.
1. Inspect tire condition; identify tire wear patterns; measure tread depth; verify tire matching (diameter and tread); inspect valve stem and cap; set tire pressure; determine needed action. (TST, MTST) P-1.
3. Check wheel mounting hardware; check wheel condition; remove and install wheel/tire assemblies (steering and drive axle); torque fasteners to manufacturer's specification using torque wrench. (IMMR, TST, MTST) P-1.

V. Electrical/Electronic Systems
B. Battery System
2. Confirm proper battery capacity for application; perform battery state-of-charge test; perform battery capacity test, determine needed action. (IMMR, TST, MTST) P-1.
3. Inspect battery, battery cables, connectors, battery boxes, mounts, and hold-downs; determine needed action. (IMMR, TST, MTST) P-1.

E. Lighting Systems
2. Test, replace, and aim headlights. (IMMR, TST, MTST) P-1.

VII. Cab
B. Instruments and Controls
2. Check operation of all accessories. (IMMR, TST, MTST) P-1.

D. Hardware
1. Check operation of wipers and washer; inspect windshield glass for cracks or discoloration; check sun visor; check seat condition, operation, and mounting; check door glass and window operation; verify operation of door and cab locks; inspect steps and grab handles; inspect mirrors, mountings, brackets, and glass; determine needed action. (IMMR, TST, MTST) P-1.

Tools and Materials: A highway truck or tractor unit in running condition, standard personal and shop tools, the following PM procedure, and a pen or pencil

Protective Clothing: Standard shop apparel including coveralls or shop coat, safety glasses, and safety footwear

PROCEDURE
Ensure that the shop LOTO is observed.

1. Complete the following form, noting whether each task has actually been performed or reviewed.

Task completed _____

Chassis Lube

Customer Name _____ RO # _____ VIN # _____ Date _____

Make _____ Model _____ Odometer _____ mi/km

Tire pressure and tread depth: *Yes / No*   **Brake Lining Check:** *Yes / No*

Inspection Code: ☑ = OK   A = Adjusted   R = Repaired   X = Replaced   N = Needs Follow-Up

N	Item #	Lube and Level	Code	Tech #
	1	Check all belts, pulleys, and tensioners for wear and adjustment.		
	2	Visually inspect engine compartment for fluid leaks (fuel, oil, and water).		
	3	Check engine oil level and adjust if needed.		
	4	Check power steering level. Inspect system for leaks.		
	5	Check clutch fluid level. If low, inspect for leaks and fill as necessary.		
	6	Check clutch adjustment.		
	7	Check clutch brake.		
	8	Check all exterior lights.		
	9	Grease chassis. Note any fittings that do not take grease.		
	10	Check driveline for excessive movement. U-joint caps must all purge grease. Check hanger bearing.		
	11	Check fluid levels in transmission and differentials.		
	12	Check oil level of front hubs. Check oil for filings. Report.		
	13	Check all wheel seals for signs of leakage.		

Optional Services

Brake Linings
Inches

Front Right	
Front Left	
2nd Axle Right	
2nd Axle Left	
3rd Axle Right	
3rd Axle Left	
4th or Lift Axle	

Front Axle

_____ psi
_____ /32

_____ psi
_____ /32

Tire Pressure and Tread Depth

Front Rear	Rear Rear
_____ psi	_____ psi
_____ /32	_____ /32
_____ psi	_____ psi
_____ /32	_____ /32
_____ psi	_____ psi
_____ /32	_____ /32
_____ psi	_____ psi
_____ /32	_____ /32

Record of Deficiencies:

Code	Tech.	Date Repaired	#	Service Procedure

Inspected by: _____   Supervisor: _____   Next Service Due: _____

FIGURE 4–3 Chassis lube checklist

STUDENT SELF-EVALUATION

Check	Level	Competency	Comments
	4	Mastered task	
	3	Competent but need further help	
	2	Needed a lot of help	
	0	Did not understand the task	

INSTRUCTOR EVALUATION

Check	Level	Competency	Comments
	4	Mastered task	
	3	Competent but needs further help	
	2	Requires more training	
	0	Unable to perform task	

JOB SHEET 4.4

Name _____ Station _____ Date _____

Perform a Summerization PM Checklist on a Linehaul Truck.

Performance Objective(s): Use the summerization form that follows to perform a pre-summer PM inspection.

ASE Education Foundation Correlation

This job sheet addresses the following ASE Education Foundation task(s):

I. Diesel Engines
   A. General
   1. Research vehicle service information, including fluid type, vehicle service history, service precautions, and technical service bulletins. (IMMR, TST, MTST) P-1.
   2. Inspect level and condition of fuel, oil, diesel exhaust fluid (DEF), and coolant. (IMMR, TST, MTST) P-1.

   D. Lubrication Systems
   2. Check engine oil level, condition, and consumption; take engine oil sample; determine needed action. (IMMR, TST, MTST) P-1.
   3. Determine proper lubricant; perform oil and filter service. (IMMR, TST, MTST) P-1.

   E. Cooling System
   1. Check engine coolant type, level, condition, and test coolant for freeze protection and additive package concentration. (IMMR, TST, MTST) P-1.
   2. Verify coolant temperature; check operation of temperature and level sensors, gauge, and/or sending unit. (IMMR) P-1.
   2. Test coolant temperature; test operation of temperature and level sensors, gauge, and/or sending unit; determine needed action. (TST, MTST) P-1.

   F. Air Induction and Exhaust Systems
   2. Check air induction system including: cooler assembly, piping, hoses, clamps, and mountings; replace air filter as needed; reset restriction indicator (if applicable). (IMMR, TST) P-1.
   2. Diagnose air induction system problems; inspect, clean, and/or replace cooler assembly, piping, hoses, clamps, and mountings; replace air filter as needed; reset restriction indicator (if applicable). (MTST) P-1.

VI. Heating, Ventilation, and Air Conditioning (HVAC)
   A. General
   3. Use appropriate electronic service tool(s) and procedures to check, record, and clear diagnostic codes; interpret digital multimeter (DMM) readings. (IMMR) P-1.
   3. Use appropriate electronic service tool(s) and procedures to diagnose problems; check, record, and clear diagnostic codes; interpret digital multimeter (DMM) readings. (TST, MTST) P-1.

   B. Refrigeration System Components
   1. Inspect A/C compressor drive belts, pulleys, and tensioners; verify proper belt alignment. (IMMR) P-1.
   1. Inspect, remove, and replace A/C compressor drive belts, pulleys, and tensioners; verify proper belt alignment. (TST, MTST) P-1.
   2. Check A/C system operation including system pressures; visually inspect A/C components for signs of leaks; check A/C monitoring system (if applicable). (IMMR, TST, MTST) P-1.

   D. Operating Systems and Related Controls
   1. Verify HVAC system blower motor operation; confirm proper air distribution; confirm proper temperature control; determine needed action. (IMMR) P-1.

Tools and Materials: A highway truck or tractor unit in running condition, standard personal and shop tools, the following maintenance inspection procedure, and a pen or pencil

Protective Clothing: Standard shop apparel including coveralls or shop coat, safety glasses, and safety footwear

PROCEDURE
Ensure that the shop LOTO is observed.

1. Complete the following form, noting whether each task has actually been performed or reviewed.

Task completed _____

Summerization

Customer Name _____ RO # _____ VIN # _____ Date _____

Make _____ Model _____ Odometer _____ mi/km

Inspection Code: ☑ = OK   A = Adjusted   R = Repaired   X = Replaced   N = Needs Follow-Up

N	Item #	Cooling System	Code	Tech #
	1	Pressure test cooling system.		
	2	Visually inspect engine compartment for fluid leaks (fuel, oil, and water).		
	3	Check hoses and clamps.		
	4	Check fan and shroud for damage.		
	5	Check all belts, pulleys, and tensioners for wear and adjustment.		
	6	Externally inspect water pump.		
	7	Check radiator condition. Inspect for obstructions, dirt, debris, etc.		
	8	Check coolant filter for leaks. Replace with correct change filter if unit requires SCA service.		
	9	Test and record coolant protection level and SCA level. Service as necessary.		
	10	Protection _____ SCA units _____		
	11	Extended Life Coolant: *Add extender if required (480,000 km, 300,000 miles, 6,000 hours).		
	12	Check thermostat operation.		
	13	Check engine fan operation.		

N	Item #	Air-Conditioning System	Code	Tech #
	14	Inspect evaporator core and heater core if combined, clean away any debris, verify drain is clear.		
	15	Check radiator, CAC and AC condenser condition, mounting, obstructions, dirt, debris, etc.		
	16	Inspect hose condition, check routing and clipping.		
	17	Inspect AC wiring harness and connections.		
	18	Inspect compressor, check for loose or broken wires, check mounting bracket and hardware.		
	19	Check heater valves (be sure heater valves are closing fully) and controls for proper operation.		
	20	Connect AC manifold gauges.		
	21	Start vehicle and operate the air-conditioning system with engine at 1,200–1,500 RPM.		
	22	Check receiver/dryer service indicator (Note: should be serviced every 2 years).		
	23	Check high- and low-pressure gauge readings to ensure proper operation.		
	24	Record running system pressure: High side _____ Low side _____		
	25	If system requires service refer to RP 413 (R-12) or RP 418 (R-134a).		
	26	Record running system pressure after service is performed: High side _____ Low side _____		
	27	Verify proper clutch operation, check for any unusual noises.		
	28	Check for proper blower speed control and measure air temperature at vent closest to the evaporator.		
	29	Record duct outlet temperature _____.		

Record of Deficiencies:

Code	Tech.	Date Repaired	#	Service Procedure

Inspected by: _____   Supervisor: _____   Next Service Due: _____

FIGURE 4–4 Summerization PM checklist

STUDENT SELF-EVALUATION

Check	Level	Competency	Comments
	4	Mastered task	
	3	Competent but need further help	
	2	Needed a lot of help	
	0	Did not understand the task	

INSTRUCTOR EVALUATION

Check	Level	Competency	Comments
	4	Mastered task	
	3	Competent but needs further help	
	2	Requires more training	
	0	Unable to perform task	

JOB SHEET 4.5

Name _____ Station _____ Date _____

Perform a Winterization Service Inspection on a Linehaul Truck.

Performance Objective(s): Use the following form to guide you through a winterization service procedure. This type of maintenance inspection is typically undertaken sometime during the fall by trucking companies operating in all but the southernmost United States.

ASE Education Foundation Correlation

This job sheet addresses the following ASE Education Foundation task(s):

I. Diesel Engines
  A. General
   1. Research vehicle service information, including fluid type, vehicle service history, service precautions, and technical service bulletins. (IMMR, TST, MTST) P-1.
   2. Inspect level and condition of fuel, oil, diesel exhaust fluid (DEF), and coolant. (IMMR, TST, MTST) P-1.

  D. Lubrication Systems
   2. Check engine oil level, condition, and consumption; take engine oil sample; determine needed action. (IMMR, TST, MTST) P-1.
   3. Determine proper lubricant; perform oil and filter service. (IMMR, TST, MTST) P-1.

  E. Cooling System
   1. Check engine coolant type, level, condition, and test coolant for freeze protection and additive package concentration. (IMMR, TST, MTST) P-1.
   2. Verify coolant temperature; check operation of temperature and level sensors, gauge, and/or sending unit. (IMMR) P-1.
   2. Test coolant temperature; test operation of temperature and level sensors, gauge, and/or sending unit; determine needed action. (TST, MTST) P-1.

  F. Air Induction and Exhaust Systems
   2. Check air induction system including: cooler assembly, piping, hoses, clamps, and mountings; replace air filter as needed; reset restriction indicator (if applicable). (IMMR, TST) P-1.
   2. Diagnose air induction system problems; inspect, clean, and/or replace cooler assembly, piping, hoses, clamps, and mountings; replace air filter as needed; reset restriction indicator (if applicable). (MTST) P-1.

II. Drive Train
  C. Transmission
   1. Inspect transmission shifter and linkage; inspect transmission mounts, insulators, and mounting bolts. (IMMR) P-1.
   1. Inspect transmission shifter and linkage; inspect and/or replace transmission mounts, insulators, and mounting bolts. (TST, MTST) P-1.

  D. Driveshaft and Universal Joints
   1. Inspect, service, and/or replace driveshafts, slip joints, yokes, drive flanges, support bearings, universal joints, boots, seals, and retaining/mounting hardware; check phasing of all shafts. (IMMR, TST, MTST) P-1.

  E. Drive Axles
   2. Check drive-axle fluid level and condition; check drive-axle filter; determine needed action. (IMMR, TST, MTST) P-1.

VI. Heating, Ventilation, and Air Conditioning (HVAC)
  D. Operating Systems and Related Controls
   1. Verify HVAC system blower motor operation; confirm proper air distribution; confirm proper temperature control; determine needed action. (IMMR, TST, MTST) P-1.

Tools and Materials: A highway truck or tractor unit in running condition, standard personal and shop tools, the following PM procedure, and a pen or pencil

Protective Clothing: Standard shop apparel including coveralls or shop coat, safety glasses, and safety footwear

PROCEDURE
Ensure that the shop LOTO is observed.

1. Complete the following form, noting whether each task has actually been performed or reviewed.

Task completed _____

Winterization Service

Customer Name _____ Work Order # _____ _____

VIN # _____ Mileage _____ km/mi (circle) Hrs. _____ Unit or Lic. # _____

Inspection Code: ✓ = OK   A = Adjusted   R = Repaired   X = Replaced   N = Needs Follow-Up

N	Item #	In Cab Inspection	Code	Tech #
	1	Check operation of windshield wipers and washers. Top up washer fluid with one jug (*"Optional"*).		
	2	Check heater blower motor performance.		
	3	Check mirror heaters.		
	4	Check condition of mirrors.		
	5	Check operation of clutch and clutch brake and report.		

N	Item #	Electrical	Code	Tech #
	6	Test alternator output.		
	7	Check tightness of the nuts that secure cables to both the starter solenoid and starter motor.		
	8	Check starter draw.		
	9	Load test batteries.		
	10	Check battery mounting. Clean battery connections and seal with protectant.		
	11	Air Starter: service filters if fitted, check operation, check starter reservoir mounting.		
	12	Check operation of all lamps and reflectors—hi and low beam.		
	13	Check trailer cord operation and condition.		

N	Item #	Engine Compartment	Code	Tech #
	14	Drain engine oil. Torque drain plug to specification.		
	15	Replace engine oil filter(s).		
	16	Replace fuel filter(s).		
	17	Service fuel/water separator as required.		
	18	Check fuel heater operation.		
	19	Service engine crankcase breather as required.		
	20	Change power steering fluid and filter if required.		
	21	Test and record coolant protection level and SCA level. Service as necessary.		
	22	Protection _____ SCA units _____		
	23	Check coolant filter for leaks. Replace with correct charge filter if unit requires SCA service.		
	24	Pressure test cooling system and check all hose clamps.		
	25	Check heater shut off taps are open, if equipped.		
	26	Check block heater, if equipped.		
	27	Check pre-heater, if equipped.		
	28	Check air filter restriction and record. Advise on condition of element.		
	29	Restriction Reading (if available) _____		
	30	Check cab filters. Report on condition.		
	31	Inspect intake piping for sealing, integrity (cracks, splitting, and or holes), alignment, and clamp tightness.		
	32	Check all belts, pulleys, and tensioners for wear and adjustment.		
	33	Check radiator condition. Inspect for obstructions, dirt, debris, etc.		
	34	Check fan and shroud for damage.		
	35	Check thermostat operation.		
	36	Check engine fan operation.		
	37	Visually inspect engine compartment for fluid leaks (fuel, oil, and water).		
	38	Check splash shields located inside wheel housings and engine compartment.		
	39	Check clutch fluid level. If low, inspect for leaks and fill as necessary.		
	40	Check condition and adjustment of hood release cables.		
	41	Check turbo, exhaust manifold, and exhaust pipes for mounting and leaks.		

FIGURE 4–5 Winterization service

Winterization Service

N	Item #	Under Truck Inspection	Code	Tech #
	42	Grease chassis. Note any fittings that do not take grease.		
	43	Check driveline for excessive movement. U-joint caps must all purge grease. Check hanger bearing.		
	44	Jack up front end and check kingpins.		
	45	Check oil level of front hubs. Check oil for filings. Report.		
	46	Check all wheel seals for signs of leakage.		
	47	Check transmission and differential oil levels. Record any leaks. Top up as needed.		
	48	Check brake adjustment by stroke length. Report.		
	49	Check brake chambers for broken springs.		
	50	Check condition of air brake hose, check for interference or chaffing.		
	51	Check clutch slave cylinder for leaks. Check condition of boot. Check yoke for wear.		
	52	Drain all air tanks. Check all drains for function and leaks. Check mounting.		
	53	Top up alcohol injector bottle, if equipped.		
	54	Service air dryer.		
	55	Check gladhand seals.		

N	Item #	Cab	Code	Tech #
	56	Check cab mounting, rubber pad, and air suspension.		
	57	Check condition of sound absorption materials affixed to hood and cab.		
	58	Check condition of steel/fiberglass body panels, bumpers, and steps.		
	59	Check tightness and condition of chassis fairings.		

Record of Deficiencies:

Code	Tech.	Date Repaired	#	Service Procedure

Service Performed by: _____   Date: _____

FIGURE 4–5 (continued)

STUDENT SELF-EVALUATION

Check	Level	Competency	Comments
	4	Mastered task	
	3	Competent but need further help	
	2	Needed a lot of help	
	0	Did not understand the task	

INSTRUCTOR EVALUATION

Check	Level	Competency	Comments
	4	Mastered task	
	3	Competent but needs further help	
	2	Requires more training	
	0	Unable to perform task	

JOB SHEET 4.6

Name _____ Station _____ Date _____

Perform a Basic Oil and Filter Change on a Truck.

Performance Objective(s): Change the oil and filters on a truck. This procedure is described as a basic oil and filter change requiring that nothing be done other than those two operations, which is unusual. A more common shop procedure is the oil and filter service in the next task sheet. This requires that you do more than just change the oil and filters, though it is described as an oil and filter service.

ASE Education Foundation Correlation

This job sheet addresses the following ASE Education Foundation task(s):

I. **Diesel Engines**
   A. **General**
      1. Research vehicle service information, including fluid type, vehicle service history, service precautions, and technical service bulletins. (IMMR, TST, MTST) P-1.
      2. Inspect level and condition of fuel, oil, diesel exhaust fluid (DEF), and coolant. (IMMR, TST, MTST) P-1.
   D. **Lubrication Systems**
      2. Check engine oil level, condition, and consumption; take engine oil sample; determine needed action. (IMMR, TST, MTST) P-1.
      3. Determine proper lubricant; perform oil and filter service. (IMMR, TST, MTST) P-1.
   E. **Cooling System**
      1. Check engine coolant type, level, condition, and test coolant for freeze protection and additive package concentration. (IMMR, TST, MTST) P-1.
      2. Verify coolant temperature; check operation of temperature and level sensors, gauge, and/or sending unit. (IMMR) P-1.
      2. Test coolant temperature; test operation of temperature and level sensors, gauge, and/or sending unit; determine needed action. (TST, MTST) P-1.
   F. **Air Induction and Exhaust Systems**
      2. Check air induction system including: cooler assembly, piping, hoses, clamps, and mountings; replace air filter as needed; reset restriction indicator. (if applicable) (IMMR, TST) P-1.
      2. Diagnose air induction system problems; inspect, clean, and/or replace cooler assembly, piping, hoses, clamps, and mountings; replace air filter as needed; reset restriction indicator. (if applicable) (MTST) P-1.

II. **Drive Train**
   D. **Driveshaft and Universal Joints**
      1. Inspect, service, and/or replace driveshafts, slip joints, yokes, drive flanges, support bearings, universal joints, boots, seals, and retaining/mounting hardware; check phasing of all shafts. (IMMR, TST, MTST) P-1.
   E. **Drive Axles**
      2. Check drive-axle fluid level and condition; check drive-axle filter; determine needed action. (IMMR, TST, MTST) P-1.

Tools and Materials: A highway truck or tractor unit in running condition, standard personal and shop tools, the following PM procedure, and a pen or pencil

Protective Clothing: Standard shop apparel including coveralls or shop coat, safety glasses, and safety footwear

PROCEDURE
Ensure that the shop LOTO is observed.

1. Complete the following form, noting whether each task has actually been performed or reviewed.

   Task completed _____

Oil and Filter Service, Basic

Customer Name _____ RO # _____ VIN # _____ Date _____

Make _____ Model _____ Odometer _____

Oil Sample Required Yes / No   **Fuel Filters** Yes / No

Inspection Code: ☑ = OK   A = Adjusted   R = Repaired   X = Replaced   N = Needs Follow-Up

N	Item #	Operation Description	Code	Tech #
	1	Drain engine oil.		
	2	Take oil sample if required.		
	3	Torque drain plug to specification.		
	4	Replace engine oil filter(s).		
	5	Refill engine crankcase.		
	6	If requested, replace fuel filter(s).		
	7	Run engine. Check for leaks.		

Record of Deficiencies:

Code	Tech.	Date Repaired	#	Service Procedure

Inspected by: _____   Next Service Due: _____

FIGURE 4–6 Basic oil and filter service

STUDENT SELF-EVALUATION

Check	Level	Competency	Comments
	4	Mastered task	
	3	Competent but need further help	
	2	Needed a lot of help	
	0	Did not understand the task	

INSTRUCTOR EVALUATION

Check	Level	Competency	Comments
	4	Mastered task	
	3	Competent but needs further help	
	2	Requires more training	
	0	Unable to perform task	

JOB SHEET 4.7

Name _____ Station _____ Date _____

Perform a Wet Service on a Truck.

Performance Objective(s): This is a typical oil and filter service procedure required by one OEM. If you compare this with the previous task sheet, you will see that much more than just the changing of the oil and filters is required for this exercise. It should be noted that the procedure outlined here is more typical than the basic oil and filter change outlined in the previous task sheet.

ASE Education Foundation Correlation

This job sheet addresses the following ASE Education Foundation task(s):

I. **Diesel Engines**
   A. **General**
      1. Research vehicle service information, including fluid type, vehicle service history, service precautions, and technical service bulletins. (IMMR, TST, MTST) P-1.
      2. Inspect level and condition of fuel, oil, diesel exhaust fluid (DEF), and coolant. (IMMR, TST, MTST) P-1.
   D. **Lubrication Systems**
      2. Check engine oil level, condition, and consumption; take engine oil sample; determine needed action. (IMMR, TST, MTST) P-1.
      3. Determine proper lubricant; perform oil and filter service. (IMMR, TST, MTST) P-1.
   E. **Cooling System**
      1. Check engine coolant type, level, condition, and test coolant for freeze protection and additive package concentration. (IMMR, TST, MTST) P-1.
      2. Verify coolant temperature; check operation of temperature and level sensors, gauge, and/or sending unit. (IMMR) P-1.
      2. Test coolant temperature; test operation of temperature and level sensors, gauge, and/or sending unit; determine needed action. (TST, MTST) P-1.
   F. **Air Induction and Exhaust Systems**
      2. Check air induction system including: cooler assembly, piping, hoses, clamps, and mountings; replace air filter as needed; reset restriction indicator (if applicable). (IMMR, TST) P-1.
      2. Diagnose air induction system problems; inspect, clean, and/or replace cooler assembly, piping, hoses, clamps, and mountings; replace air filter as needed; reset restriction indicator (if applicable). (MTST) P-1.

II. **Drive Train**
   D. **Driveshaft and Universal Joints**
      1. Inspect, service, and/or replace driveshafts, slip joints, yokes, drive flanges, support bearings, universal joints, boots, seals, and retaining/mounting hardware; check phasing of all shafts. (IMMR, TST, MTST) P-1.
   E. **Drive Axles**
      2. Check drive-axle fluid level and condition; check drive-axle filter; determine needed action. (IMMR, TST, MTST) P-1.

V. **Electrical/Electronic Systems**
   B. **Battery System**
      2. Confirm proper battery capacity for application; perform battery state-of-charge test; perform battery capacity test, determine needed action. (IMMR, TST, MTST) P-1.
      3. Inspect battery, battery cables, connectors, battery boxes, mounts, and hold-downs; determine needed action. (IMMR, TST, MTST) P-1.

E. Lighting Systems
1. Inspect for brighter-than-normal, intermittent, dim, or no-light operation; determine needed action. (IMMR) P-1.
1. Identify causes of brighter-than-normal, intermittent, dim, or no-light operation; determine needed action. (TST) P-1.
1. Diagnose causes of brighter-than-normal, intermittent, dim, or no-light operation; determine needed action. (MTST) P-1.
2. Test, replace, and aim headlights. (IMMR, TST, MTST) P-1.

Tools and Materials: A highway truck or tractor unit in running condition, standard personal and shop tools, the following PM procedure, and a pen or pencil

Protective Clothing: Standard shop apparel including coveralls or shop coat, safety glasses, and safety footwear

PROCEDURE
Ensure that the shop LOTO is observed.

1. Complete the following form, noting whether each task has actually been performed or reviewed.

    Task completed _____

Oil and Filter Service

Customer Name _____ RO # _____ VIN # _____ Date _____

Make _____ Model _____ Odometer _____ km/mi

Recalls and/or Service Programs: _____ **Oil Sample Required:** *Yes / No*

Tire pressure and tread depth: *Yes / No*   **Brake Lining Check:** *Yes / No*   **Lube and Level:** *Yes / No*

Inspection Code: ✓ = OK   A = Adjusted   R = Repaired   X = Replaced   N = Needs Follow-Up

N	Item #	Operation Description	Code	Tech #
	1	Drain engine oil. **"Take Oil Sample" if required. Torque** drain plug to specification.		
	2	Replace engine oil filter(s).		
	3	Replace fuel filter(s).		
	4	Service fuel/water separator as required.		
	5	Test and record coolant protection level and SCA level. Service as necessary. **Protection** _____ | **SCA units** _____		
	6	Check coolant filter for leaks. Replace with correct charge filter if unit requires SCA service.		
	7	Check air filter restriction and record. Advise on condition of element. Restriction reading (if available) _____		
	8	Inspect intake piping for sealing, integrity (cracks, splitting, and/or holes), alignment, and clamp tightness.		
	9	Check cab filters. Report on condition.		
	10	Check all belts, pulleys, and tensioners for wear and adjustment.		
	11	Visually inspect engine compartment for fluid leaks (fuel, oil, and water).		
	12	Check clutch fluid level. If low, inspect for leaks and fill as necessary.		
	13	Check clutch adjustment.		
	14	Check clutch brake.		
	15	Check all exterior lights.		

Optional Services

N	Item #	Lube and Level — Operation Description	Code	Tech #
	16	Grease chassis. Note any fittings that do not take grease.		
	17	Check driveline for excessive movement. U-joint caps must all purge grease. Check hanger bearing.		
	18	Check fluid levels in transmission and differentials.		
	19	Check oil level of front hubs. Check oil for filings. Report.		
	20	Check all wheel seals for signs of leakage.		

Brake Linings — Inches

Front Right	
Front Left	
2nd Axle Right	
2nd Axle Left	
3rd Axle Right	
3rd Axle Left	
4th or Left Axle	

Front Axle

_____ psi
_____ /32

_____ psi
_____ /32

Tire Pressure and Tread Depth

Front Rear	Rear Rear
_____ psi	_____ psi
_____ /32	_____ /32
_____ psi	_____ psi
_____ /32	_____ /32
_____ psi	_____ psi
_____ /32	_____ /32
_____ psi	_____ psi
_____ /32	_____ /32

Record of Deficiencies:

Code	Tech.	Date Repaired	#	Service Procedure

Inspected by: _____   Supervisor: _____   Next Service Due: _____

FIGURE 4–7 Oil and filter service

STUDENT SELF-EVALUATION

Check	Level	Competency	Comments
	4	Mastered task	
	3	Competent but need further help	
	2	Needed a lot of help	
	0	Did not understand the task	

INSTRUCTOR EVALUATION

Check	Level	Competency	Comments
	4	Mastered task	
	3	Competent but needs further help	
	2	Requires more training	
	0	Unable to perform task	

JOB SHEET 4.8

Name _____ Station _____ Date _____

Perform an OEM A-Inspection on a Truck.

Performance Objective(s): Use the following checklist to complete an A-inspection on a truck. Make sure that you complete the form in detail; for instance, when checking antifreeze protection, record the actual level of protection.

ASE Education Foundation Correlation

This job sheet addresses the following ASE Education Foundation task(s):

I. Diesel Engines
  A. General
  1. Research vehicle service information, including fluid type, vehicle service history, service precautions, and technical service bulletins. (IMMR, TST, MTST) P-1.
  D. Lubrication Systems
  2. Check engine oil level, condition, and consumption; take engine oil sample; determine needed action. (IMMR, TST, MTST) P-1.
  E. Cooling System
  1. Check engine coolant type, level, condition, and test coolant for freeze protection and additive package concentration. (IMMR, TST, MTST) P-1.
  F. Air Induction and Exhaust Systems
  2. Check air induction system including: cooler assembly, piping, hoses, clamps, and mountings; replace air filter as needed; reset restriction indicator (if applicable). (IMMR, TST) P-1.
  2. Diagnose air induction system problems; inspect, clean, and/or replace cooler assembly, piping, hoses, clamps, and mountings; replace air filter as needed; reset restriction indicator (if applicable). (MTST) P-1.

II. Drive Train
  C. Transmission
  4. Check transmission fluid level and condition; determine needed action. (IMMR, TST, MTST) P-1.
  E. Drive Axles
  1. Check for fluid leaks; inspect drive-axle housing assembly, cover plates, gaskets, seals, vent/breather, and magnetic plugs. (IMMR) P-1.
  1. Check and repair fluid leaks; inspect drive-axle housing assembly, cover plates, gaskets, seals, vent/breather, and magnetic plugs. (TST, MTST) P-1.
  2. Check drive-axle fluid level and condition; check drive-axle filter; determine needed action. (IMMR, TST, MTST) P-1.

III. Brakes
  D. Air Brakes: Parking Brake System
  1. Inspect and check parking (spring) brake chamber for leaks; determine needed action. (IMMR) P-1.
  1. Inspect, test, and/or replace parking (spring) brake chamber. (TST, MTST) P-1.
  J. Wheel Bearings
  1. Clean, inspect, lubricate, and/or replace wheel bearings and races/cups; replace seals and wear rings; inspect spindle/tube; inspect and replace retaining hardware; adjust wheel bearings; check hub assembly fluid level and condition; verify end play with dial indicator method. (IMMR, TST, MTST) P-1.

IV. Suspension and Steering
  C. Steering Pump and Gear Units
  1. Check power steering pump and gear operation, mountings, lines, and hoses; check fluid level and condition; service filter; inspect system for leaks. (IMMR, TST, MTST) P-1.

G. Wheels and Tires
1. Inspect tire condition; identify tire wear patterns; measure tread depth; verify tire matching (diameter and tread); inspect valve stem and cap; set tire pressure. (IMMR) P-1.
1. Inspect tire condition; identify tire wear patterns; measure tread depth; verify tire matching (diameter and tread); inspect valve stem and cap; set tire pressure; determine needed action. (TST, MTST) P-1.
3. Check wheel mounting hardware; check wheel condition; remove and install wheel/tire assemblies (steering and drive axle); torque fasteners to manufacturer's specification using torque wrench. (IMMR, TST, MTST) P-1.

V. Electrical/Electronic Systems
E. Lighting Systems
2. Test, replace, and aim headlights. (IMMR, TST, MTST) P-1.

VII. Cab
B. Instruments and Controls
2. Check operation of all accessories. (IMMR, TST, MTST) P-1.

Tools and Materials: A highway truck or tractor unit in running condition, standard personal and shop tools, the following PM procedure, and a pen or pencil

Protective Clothing: Standard shop apparel including coveralls or shop coat, safety glasses, and safety footwear

PROCEDURE
Ensure that the shop LOTO is observed.

1. Complete the following form, noting whether each task has actually been performed or reviewed.

Task completed _____

A Service

Customer Name _____ RO # _____ VIN # _____ Date _____

Make _____ Model _____ Odometer _____ km/mi

Recalls and/or Service Programs: _____ **Oil Sample Required:** *Yes / No*

Inspection Code: ☑ = OK   A = Adjusted   R = Repaired   X = Replaced   N = Needs Follow-Up

N	Item #	In Cab Inspection	Code	Tech #
	1	Check and record ECU fault codes.		
	2	Check operation of windshield wipers, washers, and condition of windshield and visors.		
	3	Check operation of all warning lamps and alarms.		
	4	Check operation of tractor protection valve.		
	5	Check operation of air horn _____ and electric horn _____ .		
	6	Check operation of clutch, clutch brake, free travel, and pedal pad.		
	7	Inspect mirror mountings, brackets, and glass.		

N	Item #	Outside Cab	Code	Tech #
	8	Check headlamps, turn signals, fog, clearance, and brake lights.		
	9	Check trailer cord operation and condition.		
	10	Check trailer air lines and gladhand seals.		
	11	Check condition of mud flaps, hangers, and fender mounting.		
	12	Check muffler and exhaust stack mounting. Check for leaks.		
	13	Check fifth wheel adjustment, mounting, and slide operation; lubricate if needed.		

N	Item #	Engine Compartment	Code	Tech #
	14	Check engine oil level and adjust if needed.		
	15	Drain and clean fuel/water separator as necessary.		
	16	Test and **record** coolant protection level and SCA level. Service as necessary. **Protection** _____ SCA units _____		
	17	Check air filter restriction and record. Advise on condition of element. Restriction Reading (if available) _____		
	18	Inspect intake piping for sealing, integrity (cracks, splitting, and/or holes), alignment, and clamp tightness.		
	19	Check all belts, pulleys, and tensioners for wear and adjustment.		
	20	Visually inspect engine compartment for fluid leaks (fuel, oil, and water).		
	21	Check power steering level. Inspect system for leaks.		
	22	Check clutch fluid level. If low, inspect for leaks and fill as necessary.		
	23	Top up washer fluid with one jug (*"Optional"*). Any remainder to be put in cab.		
	24	Check mounting of alternator and air-conditioning compressor. Check for broken or loose hardware.		
	25	Check turbo, exhaust manifold, and exhaust pipes for mounting and leaks.		

N	Item #	Under Truck Inspection	Code	Tech #
	26	Grease chassis. Note any fittings that do not take grease.		
	27	Check driveline for excessive movement. U-joint caps must all purge grease. Check hanger bearing.		
	28	Check oil level of front hubs. Check oil for filings. Report.		
	29	Check all wheel seals for signs of leakage.		
	30	Check all brakes for lining. Note lining thickness below.		
	31	Check brake chambers for broken springs, rubbing lines.		
	32	Check brake adjustment by stroke length. Report.		
	33	Check transmission and differential oil levels, inspect and clean breathers, check for leaks.		
	34	Check clutch slave cylinder for leaks. Check condition of boot. Check yoke for wear.		
	35	Check transmission cooler mounting. Inspect cooler hose condition.		
	36	Check and adjust tire air pressures, measure tread depth. Note readings and list any defects found.		
	37	Install "Next Service Due" decal.		
	38	Park unit. Check for grease and oil marks. Clean as necessary. Lock unit.		

FIGURE 4–8 A-service inspection

Brake Linings

Inches

Front Right	
Front Left	
2nd Axle Right	
2nd Axle Left	
3rd Axle Right	
3rd Axle Left	
4th or Lift Axle	

Tire Pressure and Tread Depth

Front Axle

_____ psi
_____ /32

_____ psi
_____ /32

Front Rear	Rear Rear
_____ psi	_____ psi
_____ /32	_____ /32
_____ psi	_____ psi
_____ /32	_____ /32
_____ psi	_____ psi
_____ /32	_____ /32
_____ psi	_____ psi
_____ /32	_____ /32

Record of Deficiencies:

Code	Tech.	Date Repaired	#	Service Procedure

Inspected by: _____ Supervisor: _____ Next Service Due: _____

FIGURE 4–8 (continued)

STUDENT SELF-EVALUATION

Check	Level	Competency	Comments
	4	Mastered task	
	3	Competent but need further help	
	2	Needed a lot of help	
	0	Did not understand the task	

INSTRUCTOR EVALUATION

Check	Level	Competency	Comments
	4	Mastered task	
	3	Competent but needs further help	
	2	Requires more training	
	0	Unable to perform task	

JOB SHEET 4.9

Name _____ Station _____ Date _____

Perform a B-Inspection on a Truck.

Performance Objective(s): Use the following form to perform an OEM B-inspection on a truck. The objective is to familiarize yourself with exactly what is required of a typical B-grade PM inspection, but you may not be able to perform all of the tasks as they appear on the checklist due to the cost factors involved. Replacing engine oil and filters and undertaking oil analyses are expensive procedures in a training environment.

ASE Education Foundation Correlation

This job sheet addresses the following ASE Education Foundation task(s):

I. Diesel Engines
  A. General
  1. Research vehicle service information, including fluid type, vehicle service history, service precautions, and technical service bulletins. (IMMR, TST, MTST) P-1.

  D. Lubrication Systems
  2. Check engine oil level, condition, and consumption; take engine oil sample; determine needed action. (IMMR, TST, MTST) P-1.

  E. Cooling System
  1. Check engine coolant type, level, condition, and test coolant for freeze protection and additive package concentration. (IMMR, TST, MTST) P-1.

  F. Air Induction and Exhaust Systems
  2. Check air induction system including: cooler assembly, piping, hoses, clamps, and mountings; replace air filter as needed; reset restriction indicator (if applicable). (IMMR, TST) P-1.
  2. Diagnose air induction system problems; inspect, clean, and/or replace cooler assembly, piping, hoses, clamps, and mountings; replace air filter as needed; reset restriction indicator (if applicable). (MTST) P-1.

  G. Fuel System
  2. Inspect fuel tanks, vents, caps, mounts, valves, screens, crossover system, hoses, lines, and fittings; determine needed action. (IMMR, TST, MTST) P-1.
  4. Replace fuel filter; prime and bleed fuel system. (IMMR, TST, MTST) P-1.

II. Drive Train
  B. Clutch
  2. Inspect clutch master cylinder fluid level; check clutch master cylinder, slave cylinder, lines, and hoses for leaks and damage; determine needed action. (IMMR, TST, MTST) P-1.

  C. Transmission
  4. Check transmission fluid level and condition; determine needed action. (IMMR, TST, MTST) P-1.
  5. Inspect transmission breather; inspect transmission oil filters, coolers, and related components; determine needed action. (IMMR, TST, MTST) P-2.

  E. Drive Axles
  1. Check for fluid leaks; inspect drive-axle housing assembly, cover plates, gaskets, seals, vent/breather, and magnetic plugs. (IMMR) P-1.
  1. Check and repair fluid leaks; inspect drive-axle housing assembly, cover plates, gaskets, seals, vent/breather, and magnetic plugs. (TST, MTST) P-1.
  2. Check drive-axle fluid level and condition; check drive-axle filter; determine needed action. (IMMR, TST, MTST) P-1.

III. Brakes
  A. General
  2. Identify brake system components and configurations (including air and hydraulic systems, parking brake, power assist, and vehicle dynamic brake systems). (IMMR, TST, MTST) P-1.

D. Air Brakes: Parking Brake System
1. Inspect and check parking (spring) brake chamber for leaks; determine needed action. (IMMR) P-1.
1. Inspect, test, and/or replace parking (spring) brake chamber. (TST, MTST) P-1.

E. Hydraulic Brakes: Hydraulic System
2. Inspect hydraulic brake system components for leaks and damage. (IMMR) P-1.
2. Inspect hydraulic brake system for leaks and damage; test, repair, and/or replace hydraulic brake system components. (TST, MTST) P-1.

J. Wheel Bearings
1. Clean, inspect, lubricate, and/or replace wheel bearings and races/cups; replace seals and wear rings; inspect spindle/tube; inspect and replace retaining hardware; adjust wheel bearings; check hub assembly fluid level and condition; verify end play with dial indicator method. (IMMR, TST, MTST) P-1.

IV. Suspension and Steering
C. Steering Pump and Gear Units
1. Check power steering pump and gear operation, mountings, lines, and hoses; check fluid level and condition; service filter; inspect system for leaks. (IMMR, TST, MTST) P-1.

G. Wheels and Tires
1. Inspect tire condition; identify tire wear patterns; measure tread depth; verify tire matching (diameter and tread); inspect valve stem and cap; set tire pressure. (IMMR) P-1.
1. Inspect tire condition; identify tire wear patterns; measure tread depth; verify tire matching (diameter and tread); inspect valve stem and cap; set tire pressure; determine needed action. (TST, MTST) P-1.
3. Check wheel mounting hardware; check wheel condition; remove and install wheel/tire assemblies (steering and drive axle); torque fasteners to manufacturer's specification using torque wrench. (IMMR, TST, MTST) P-1.

V. Electrical/Electronic Systems
B. Battery System
2. Confirm proper battery capacity for application; perform battery state-of-charge test; perform battery capacity test, determine needed action. (IMMR, TST, MTST) P-1.
3. Inspect battery, battery cables, connectors, battery boxes, mounts, and hold-downs; determine needed action. (IMMR, TST, MTST) P-1.

E. Lighting Systems
2. Test, replace, and aim headlights. (IMMR, TST, MTST) P-1.

VII. Cab
B. Instruments and Controls
2. Check operation of all accessories. (IMMR, TST, MTST) P-1.

D. Hardware
1. Check operation of wipers and washer; inspect windshield glass for cracks or discoloration; check sun visor; check seat condition, operation, and mounting; check door glass and window operation; verify operation of door and cab locks; inspect steps and grab handles; inspect mirrors, mountings, brackets, and glass; determine needed action. (IMMR, TST, MTST) P-1.

Tools and Materials: A highway truck or tractor unit in running condition, standard personal and shop tools, the following PM procedure, and a pen or pencil

Protective Clothing: Standard shop apparel including coveralls or shop coat, safety glasses, and safety footwear

PROCEDURE
Ensure that the shop LOTO is observed.

1. Complete the following form, noting whether each task has actually been performed or reviewed.

Task completed _____

B Service

Customer Name _____ RO # _____ VIN # _____ Date _____

Make _____ Model _____ Odometer _____ km/mi

Recalls and/or Service Programs: _____ **Oil Sample Required:** *Yes / No*

Inspection Code: ✓ = OK   A = Adjusted   R = Repaired   X = Replaced   N = Needs Follow-Up

N	Item #	In Cab Inspection	Code	Tech #
	1	Check and record ECU fault codes.		
	2	Check seat mounting, seat belt operation, and sleeper restraints.		
	3	Check operation of windshield wipers, washers, and condition of windshield and visors.		
	4	Check operation of parking brake. Check for air leaks, service brakes applied and released.		
	5	Check operation of all warning lamps and alarms.		
	6	Check and record cut in _____ and cut out _____ pressure of air system.		
	7	Check operation of tractor protection valve.		
	8	Check operation of air horn _____ and electric horn _____ .		
	9	Check operation of clutch, clutch brake, free travel, and pedal pad.		
	10	Inspect mirror mountings, brackets, and glass.		

N	Item #	Outside Cab	Code	Tech #
	11	Check headlamps, turn signals, fog, clearance, and brake lights.		
	12	Check trailer cord operation and condition.		
	13	Check trailer air lines and gladhand seals.		
	14	Check condition of mud flaps, hangers, and fender mounting.		
	15	Check muffler and exhaust stack mounting. Check for leaks.		
	16	Check cab suspension components.		
	17	Check fifth wheel adjustment, mounting, and slide operation.		
	18	Check condition and mounting of rear shocks and air bags for any defects.		
	19	Check fuel tanks, mountings, lines, and caps.		
	20	Check battery mounting. Clean battery connections and seal.		

N	Item #	Engine Compartment	Code	Tech #
	21	Drain engine oil. "Take Oil Sample" if required as per kit instructions. Torque drain plug to spec.		
	22	Replace engine oil filter(s).		
	23	Replace fuel filter(s). Clean and/or replace prepump filter screen, if fitted.		
	24	Drain and clean fuel/water separator as necessary.		
	25	Test and record coolant protection level and SCA level. Service as necessary. Protection _____ SCA units _____		
	26	Check coolant filter for leaks. Replace with correct charge filter if unit requires SCA service.		
	27	Check air filter restriction and record. Advise on condition of element. Restriction Reading (if available) _____		
	28	Inspect intake piping for sealing, integrity (cracks, splitting, and/or holes), alignment, and clamp tightness.		
	29	Check cab filters. Report on condition.		
	30	Check all belts, pulleys, and tensioners for wear and adjustment.		
	31	Check radiator, CAC and AC condenser condition, mounting, obstructions, dirt, debris. etc.		
	32	Check fan and shroud for damage.		
	33	Check thermostat operation.		
	34	Check engine fan operation.		
	35	Visually inspect engine compartment for fluid leaks (fuel, oil, and water).		
	36	Check power steering level. Inspect system for leaks.		
	37	Check opening operation of hood shocks and mounting.		
	38	Check hood insulation, if equipped.		
	39	Check drag link, tie-rod ends, and steering shaft for wear or looseness.		
	40	Check clutch fluid level. If low, inspect for leaks and fill as necessary.		
	41	Top up washer fluid with one jug ("Optional"). Any remainder to be put in cab.		

FIGURE 4–9 B-inspection form

B Service

N	Item #		Code	Tech #
	42	Check air to air pipes, hoses, and clamps.		
	43	Check condition and adjustment of hood release cables.		
	44	Check mounting of alternator and air-conditioning compressor. Check for broken or loose hardware.		
	45	Check turbo, exhaust manifold, and exhaust pipes for mounting and leaks.		
	46	Check starter mounting, cable connections for corrosion and tightness.		
N	**Item #**	**Under Truck Inspection**	**Code**	**Tech #**
	47	Grease chassis. Note any fittings that do not take grease.		
	48	Check driveline for excessive movement. U-joint caps must all purge grease. Check hanger bearing.		
	49	Jack up front end and check kingpins.		
	50	Check front wheel bearing end play.		
	51	Check oil level of front hubs. Check oil for filings. Report.		
	52	Check all wheel seals for signs of leakage.		
	53	Check all brakes for lining. Note lining thickness below.		
	54	Check brake chambers for broken springs, rubbing lines.		
	55	Check brake adjustment by stroke length. Report.		
	56	Check transmission and differential oil levels. Inspect and clean breathers. Check for leaks.		
	57	Check clutch slave cylinder for leaks. Check condition of boot. Check yoke for wear.		
	58	Check transmission cooler mounting. Inspect cooler hose condition.		
	59	Inspect transmission mounts for wear or cracking.		
	60	Check suspension U-bolts and saddles. Look for any shift in components.		
	61	Check suspension torque rods for bushing wear.		
	62	Check air ride linkage for wear.		
	63	Check all rims and fasteners for cracks.		
	64	Torque all wheels.		
	65	Check and adjust tire air pressures, measure tread depth. Note readings and list any defects founds.		
	66	Run unit. Recheck oil level. Wipe off oil and fuel filters. Check for leaks.		
	67	Install "Next Service Due" decal.		
	68	Park unit. Check for grease and oil marks. Clean as necessary. Lock unit.		

Brake Linings
Inches

Front Right	
Front Left	
2nd Axle Right	
2nd Axle Left	
3rd Axle Right	
3rd Axle Left	
4th or Lift Axle	

Front Axle

_____ psi
_____ /32

_____ psi
_____ /32

Tire Pressure and Tread Depth

Front Rear	Rear Rear
_____ psi	_____ psi
_____ /32	_____ /32
_____ psi	_____ psi
_____ /32	_____ /32
_____ psi	_____ psi
_____ /32	_____ /32
_____ psi	_____ psi
_____ /32	_____ /32

Record of Deficiencies:

Code	Tech.	Date Repaired	#	Service Procedure

Inspected by: _____ Supervisor: _____ Next Service Due: _____

FIGURE 4-9 (continued)

STUDENT SELF-EVALUATION

Check	Level	Competency	Comments
	4	Mastered task	
	3	Competent but need further help	
	2	Needed a lot of help	
	0	Did not understand the task	

INSTRUCTOR EVALUATION

Check	Level	Competency	Comments
	4	Mastered task	
	3	Competent but needs further help	
	2	Requires more training	
	0	Unable to perform task	

JOB SHEET 4.10

Name _____ Station _____ Date _____

Perform a Dry Van Trailer Service.

Performance Objective(s): Perform a full service on a dry van trailer. You should note that trailers generate more OOS tickets than the tractors that haul them, and part of the reason is shortcomings in PM practice.

ASE Education Foundation Correlation

This job sheet addresses the following ASE Education Foundation task(s):

III. Brakes
A. General
2. Identify brake system components and configurations (including air and hydraulic systems, parking brake, power assist, and vehicle dynamic brake systems). (IMMR, TST, MTST) P-1.

C. Air Brakes: Mechanical/Foundation Brake Systems
1. Inspect service brake chambers, diaphragms, clamps, springs, pushrods, clevises, and mounting brackets; determine needed action. (IMMR) P-1.
1. Inspect and test service brake chambers, diaphragms, clamps, springs, pushrods, clevises, and mounting brackets; determine needed action. (TST) P-1.
1. Inspect, test, repair, and/or replace service brake chambers, diaphragms, clamps, springs, pushrods, clevises, and mounting brackets; determine needed action. (MTST) P-1.
2. Identify slack adjuster type; inspect slack adjusters; determine needed action. (IMMR) P-1.
2. Identify slack adjuster type; inspect slack adjusters; perform needed action. (TST, MTST) P-1.
3. Check camshafts (S-cams), tubes, rollers, bushings, seals, spacers, retainers, brake spiders, shields, anchor pins, and springs; determine needed action. (IMMR) P-1.
3. Check camshafts (S-cams), tubes, rollers, bushings, seals, spacers, retainers, brake spiders, shields, anchor pins, and springs; perform needed action. (TST, MTST) P-1.
6. Remove brake drum; clean and inspect brake drum and mounting surface; measure brake drum diameter; measure brake lining thickness; inspect brake lining condition; determine needed action. (IMMR, TST, MTST) P-1.

D. Air Brakes: Parking Brake System
1. Inspect and check parking (spring) brake chamber for leaks; determine needed action. (IMMR) P-1.
1. Inspect, test, and/or replace parking (spring) brake chamber. (TST, MTST) P-1.
2. Inspect and test parking (spring) brake check valves, lines, hoses, and fittings; determine needed action. (IMMR) P-1.
2. Inspect, test, and/or replace parking (spring) brake check valves, lines, hoses, and fittings. (TST, MTST) P-1.
3. Inspect and test parking (spring) brake application and release valve; determine needed action. (IMMR) P-1.
3. Inspect, test, and/or replace parking (spring) brake application and release valve. (TST, MTST) P-1.
4. Manually release (cage) and reset (uncage) parking (spring) brakes. (IMMR, TST, MTST) P-1.

J. Wheel Bearings
1. Clean, inspect, lubricate, and/or replace wheel bearings and races/cups; replace seals and wear rings; inspect spindle/tube; inspect and replace retaining hardware; adjust wheel bearings; check hub assembly fluid level and condition; verify end play with dial indicator method. (IMMR, TST, MTST) P-1.
2. Identify, inspect, and/or replace unitized/preset hub bearing assemblies. (IMMR, TST, MTST) P-2.

IV. Suspension and Steering
E. Suspension Systems
1. Inspect shock absorbers, bushings, brackets, and mounts; determine needed action. (IMMR) P-1.
1. Inspect, service, repair, and/or replace shock absorbers, bushings, brackets, and mounts. (TST, MTST) P-1.
2. Inspect leaf springs, center bolts, clips, pins, bushings, shackles, U-bolts, insulators, brackets, and mounts; determine needed action. (IMMR) P-1.

2. Inspect, repair, and/or replace leaf springs, center bolts, clips, pins, bushings, shackles, U-bolts, insulators, brackets, and mounts. (TST, MTST) P-1.
   5. Inspect and test air suspension pressure regulator and height control valves, lines, hoses, dump valves, and fittings; check and record ride height. (IMMR) P-1.

G. Wheels and Tires
   1. Inspect tire condition; identify tire wear patterns; measure tread depth; verify tire matching (diameter and tread); inspect valve stem and cap; set tire pressure. (IMMR) P-1.
   1. Inspect tire condition; identify tire wear patterns; measure tread depth; verify tire matching (diameter and tread); inspect valve stem and cap; set tire pressure; determine needed action. (TST, MTST) P-1.
   3. Check wheel mounting hardware; check wheel condition; remove and install wheel/tire assemblies (steering and drive axle); torque fasteners to manufacturer's specification using torque wrench. (IMMR, TST, MTST) P-1.

H. Frame and Coupling Devices
   1. Inspect, service, and/or adjust fifth wheel, pivot pins, bushings, locking mechanisms, mounting hardware, air lines, and fittings. (IMMR, TST, MTST) P-1.
   2. Inspect frame and frame members for cracks, breaks, corrosion, distortion, elongated holes, looseness, and damage; determine needed action. (IMMR, TST, MTST) P-1.
   4. Check pintle hook and mounting (if applicable). (IMMR) P-1.
   4. Inspect, repair, or replace pintle hooks and draw bars (if applicable). (TST, MTST) P-2.

VII. Cab
B. Instruments and Controls
   2. Check operation of all accessories. (IMMR, TST, MTST) P-1.

Tools and Materials: A highway truck or tractor unit in running condition, standard personal and shop tools, the following PM procedure, and a pen or pencil

Protective Clothing: Standard shop apparel including coveralls or shop coat, safety glasses, and safety footwear

PROCEDURE
Ensure that the shop LOTO is observed.

   1. Complete the following form, noting whether each task has actually been performed or reviewed.

Task completed _____

Trailer Service (Dry Van)

Customer Name _____ RO # _____ VIN # _____ Date _____

Make _____ Model _____ Hub Odometer _____ km/mi

Inspection Code: ☑ = OK   A = Adjusted   R = Repaired   X = Replaced   N = Needs Follow-Up

N	Item #	Exterior Visual Inspection	Code	Tech #
	1	Inspect SAE J560 seven-pin receptacle.		
	2	Inspect gladhands.		
	3	Check operation of all lights.		
	4	Inspect light lenses and reflectors.		
	5	Inspect **rear** door(s), handle(s), hinges, lock(s), tie backs, and cables.		
	6	Operate **rear** door(s) and inspect door seals and tracks.		
	7	Check license plate and mounting.		
	8	Inspect **side** door(s), handle(s), hinges, lock(s), tie backs, and cables.		
	9	Operate **side** door(s) and inspect door seals and tracks.		
	10	Inspect rear bumper.		
	11	Record all physical damage (record under deficiencies).		
		Under Carriage & Suspension		
	12	Inspect upper coupler plate and king pin.		
	13	Inspect cross members and substructures.		
	14	Inspect wiring harness(es) and air lines.		
	15	Inspect trailer landing gear and test operation.		
	16	Inspect springs, hangers, torque rods, radius rods, and bushings.		
	17	Inspect tandem slide and lock mechanism (if trailer is so equipped).		
	18	Check air bags for leaks and damage, levelling valve and linkage (air ride if equipped).		
	19	Inspect spare tire carrier.		
	20	Inspect mud flaps and brackets.		
	21	Inspect rear pintle hitch.		
	22	Inspect SAE J560 seven-pin connector receptacle and gladhands **(rear)**.		
		Brakes		
	23	Inspect brake linings and drums.		
	24	Record lining thickness below.		
	25	Check air brake system for leaks.		
	26	Adjust all brakes (manual).		
	27	Check operation of automatic slacks (if so equipped).		
	28	Drain air tanks.		
	29	Inspect air tank and brake chamber mounting.		
		Hubs		
	30	Check wheel bearings.		
	31	Check hub caps for leaks.		
	32	Check lube levels.		
		Tires and Wheels		
	33	Check for irregular wear, cuts, and sidewall damage.		
	34	Inspect and record tread depth below.		
	35	Check and adjust tire pressures, measure tread depth. Note readings and list any defects found.		
	36	Inspect for loose lugs, cracked, damaged, or slipped wheels.		

FIGURE 4–10 Trailer dry van service

Trailer Service (Dry Van)

Brake Linings

	Inches
1st Axle Right	
1st Axle Left	
2nd Axle Right	
2nd Axle Left	
3rd Axle Right	
3rd Axle Left	
4th Axle Right	
4th Axle Left	

Tire Pressure and Tread Depth

1st Axle	2nd Axle	3rd Axle	4th Axle
_____ psi	_____ psi	_____ psi	_____ psi
_____ /32	_____ /32	_____ /32	_____ /32
_____ psi	_____ psi	_____ psi	_____ psi
_____ /32	_____ /32	_____ /32	_____ /32
_____ psi	_____ psi	_____ psi	_____ psi
_____ /32	_____ /32	_____ /32	_____ /32
_____ psi	_____ psi	_____ psi	_____ psi
_____ /32	_____ /32	_____ /32	_____ /32

Record of Deficiencies:

Code	Tech.	Date Repaired	#	Service Procedure

Inspected by: _____ Supervisor: _____ Next Service Due: _____

FIGURE 4–10 (continued)

STUDENT SELF-EVALUATION

Check	Level	Competency	Comments
	4	Mastered task	
	3	Competent but need further help	
	2	Needed a lot of help	
	0	Did not understand the task	

INSTRUCTOR EVALUATION

Check	Level	Competency	Comments
	4	Mastered task	
	3	Competent but needs further help	
	2	Requires more training	
	0	Unable to perform task	

JOB SHEET 4.11

Name _____ Station _____ Date _____

Perform a Pre-delivery Inspection (PDI) on a Truck.

Performance Objective(s): Perform a pre-delivery inspection (PDI) on a truck. All manufacturers require at least one PDI to be undertaken, usually by the dealership, prior to delivery of a new truck to a customer. You should also note that most fleets repeat every step on a PDI after taking delivery and before putting the truck in service. Nothing can do more damage to OEM reputation than when a customer can identify simple shortcomings in a new vehicle that should have been picked up on a PDI. You do not need a new truck to perform this checklist; the idea is to give you a sense of what most OEMs require.

ASE Education Foundation Correlation

This job sheet addresses the following ASE Education Foundation task(s):

I. **Diesel Engines**
   A. General
   1. Research vehicle service information, including fluid type, vehicle service history, service precautions, and technical service bulletins. (IMMR, TST, MTST) P-1.

   D. Lubrication Systems
   2. Check engine oil level, condition, and consumption; take engine oil sample; determine needed action. (IMMR, TST, MTST) P-1.

   E. Cooling System
   1. Check engine coolant type, level, condition, and test coolant for freeze protection and additive package concentration. (IMMR, TST, MTST) P-1.

   F. Air Induction and Exhaust Systems
   2. Check air induction system including: cooler assembly, piping, hoses, clamps, and mountings; replace air filter as needed; reset restriction indicator (if applicable). (IMMR, TST) P-1.
   2. Diagnose air induction system problems; inspect, clean, and/or replace cooler assembly, piping, hoses, clamps, and mountings; replace air filter as needed; reset restriction indicator (if applicable). (MTST) P-1.

   G. Fuel System
   2. Inspect fuel tanks, vents, caps, mounts, valves, screens, crossover system, hoses, lines, and fittings; determine needed action. (IMMR, TST, MTST) P-1.
   4. Replace fuel filter; prime and bleed fuel system. (IMMR, TST, MTST) P-1.

II. **Drive Train**
   B. Clutch
   2. Inspect clutch master cylinder fluid level; check clutch master cylinder, slave cylinder, lines, and hoses for leaks and damage; determine needed action. (IMMR, TST, MTST) P-1.

   C. Transmission
   4. Check transmission fluid level and condition; determine needed action. (IMMR, TST, MTST) P-1.
   5. Inspect transmission breather; inspect transmission oil filters, coolers, and related components; determine needed action. (IMMR, TST, MTST) P-2.

   D. Driveshaft and Universal Joints
   1. Inspect, service, and/or replace driveshafts, slip joints, yokes, drive flanges, support bearings, universal joints, boots, seals, and retaining/mounting hardware; check phasing of all shafts. (IMMR, TST, MTST) P-1.

   E. Drive Axles
   1. Check for fluid leaks; inspect drive-axle housing assembly, cover plates, gaskets, seals, vent/breather, and magnetic plugs. (IMMR) P-1.
   1. Check and repair fluid leaks; inspect drive axle housing assembly, cover plates, gaskets, seals, vent/breather, and magnetic plugs. (TST, MTST) P-1.
   2. Check drive-axle fluid level and condition; check drive-axle filter; determine needed action. (IMMR, TST, MTST) P-1

III. Brakes
J. Wheel Bearings
1. Clean, inspect, lubricate, and/or replace wheel bearings and races/cups; replace seals and wear rings; inspect spindle/tube; inspect and replace retaining hardware; adjust wheel bearings; check hub assembly fluid level and condition; verify end play with dial indicator method. (IMMR, TST, MTST) P-1.

IV. Suspension and Steering
C. Steering Pump and Gear Units
1. Check power steering pump and gear operation, mountings, lines, and hoses; check fluid level and condition; service filter; inspect system for leaks. (IMMR, TST, MTST) P-1.

G. Wheels and Tires
3. Check wheel mounting hardware; check wheel condition; remove and install wheel/tire assemblies (steering and drive axle); torque fasteners to manufacturer's specification using torque wrench. (IMMR, TST, MTST) P-1.

V. Electrical/Electronic Systems
B. Battery System
2. Confirm proper battery capacity for application; perform battery state-of-charge test; perform battery capacity test; determine needed action. (IMMR, TST, MTST) P-1.

E. Lighting Systems
2. Test, replace, and aim headlights. (IMMR, TST, MTST) P-1.

VII. Cab
B. Instruments and Controls
2. Check operation of all accessories. (IMMR, TST, MTST) P-1.

D. Hardware
1. Check operation of wipers and washer; inspect windshield glass for cracks or discoloration; check sun visor; check seat condition, operation, and mounting; check door glass and window operation; verify operation of door and cab locks; inspect steps and grab handles; inspect mirrors, mountings, brackets, and glass; determine needed action. (IMMR, TST, MTST) P-1.

Tools and Materials: A highway truck or tractor unit in running condition, standard personal and shop tools, the following PM procedure, and a pen or pencil

Protective Clothing: Standard shop apparel including coveralls or shop coat, safety glasses, and safety footwear

PROCEDURE
Ensure that the shop LOTO is observed.

1. Complete the following form, noting whether each task has actually been performed or reviewed.

Task completed _____

New Vehicle Pre-Delivery Inspection (PDI)

OEM _____	☐ Complete Vehicle	Serial Number
VIN _____	☐ Chassis Only	
Model year _____	☐ Cab/Chassis	
Dealer Code		Repair Order
Inspection Date / /	Odometer mi	kms

- Complete all sections and fill in **code**: **X = Okay   A = Adjusted   N/A = Not Applicable**. Instructions are included at each inspection point.
- Note that top-off fluids will not be reimbursed unless component is found dry from the factory.
- **IMPORTANT:** Before starting the engine, check for correct oil and coolant levels.

CODE	CODES: X = Okay A = Adjusted N/A = Not Applicable
A. Cab Interior/Exterior - Operational Check	
Low Air Warning System	Confirm that the low-air-pressure warning system operates correctly.
Windshield Wipers and Washer	Check the windshield reservoir level. Adjust to the full mark, if necessary. Make sure the wipers and washers operate correctly.
Electric and Air Horn(s)	Confirm that the horn(s) operate.
Safety Belts and Seat Control Operation and Mounting	Inspect to make sure the seats are fully adjustable, the mounting fasteners are tight, and the seat belts operate correctly.
Hydraulic Brake Booster Warning System/Electric Motor Backup	Check for correct operation (for vehicles equipped with hydraulic brakes).
Static Air Pressure Test	1. With parking brake set, transmission in neutral, and tires chocked, run engine until the air pressure reads 120 psi (827 kPa). 2. Turn the engine off. The pressure should not drop more than 10 psi (69 kPa) within 5 minutes. 3. Tighten air line fitting as required.
Lights (Headlights, Road lights, Daytime Running, Marker, Turn Signal, Stop, and Tail)	Confirm that all headlights, road lights, daytime running lights, marker lights, turn signals, stoplights, and taillights operate properly and adjust as necessary.
B. Under Hood or Cab - Visual Inspection *(tighten or adjust only)*	
Engine Oil Leaks	Check for oil leaks and tighten all fittings and fasteners, as necessary.
Engine Oil Level; Check levels	Check engine oil level and add oil if necessary. • To deliver the vehicle to a customer, the oil level should be halfway between the low and high marks. Do not add oil if the oil level is above the midway point. Do not fill if beyond the high mark.
Coolant and Heater Hoses	Inspect all radiator and heater hose connections for leakage and tighten connections, as necessary. • If a hose routing deviation is found as part of the PDI, refer to the OEM service literature.
Coolant Level; Check	Check coolant level in the radiator and fill to correct level or to the minimum antifreeze concentration, as needed. • The concentration must provide protection to –30°F (–34°C) or below.
Fuel Leaks	Inspect for fuel leaks and tighten fittings, as necessary.
Power Steering Fluid	Inspect for power steering fluid leaks and tighten fittings, as necessary. Check reservoir fluid level. Add fluid, as necessary.

FIGURE 4–11 Pre-delivery inspection (PDI) worksheet

CODE	CODES: X = Okay A = Adjusted	N/A = Not Applicable
	Automatic Transmission; Run cold and hot checks	Check fluid level and add fluid, if needed.
	Clutch Hydraulic Fluid in units equipped with hydraulic clutch.	Check the fluid level and add fluid, if needed.
	Brake Hydraulic Fluid in units equipped with hydraulic brakes.	Check the fluid level and add fluid, if needed.
	HVAC System Operation	Check HVAC operation. Inspect for leaks around condenser, A/C lines, and fittings.
C. Chassis - Inspection		
	Drive-Axle Flange Fasteners	Tighten drive-axle flange fasteners to the correct torque specs.
	Steering Linkage	Insect for Torque Seal on the steering-driveline-yoke pinch bolts, Pitman-arm pinch bolt, drag link to steering-arm ball stud, and the tie-rod ends clamp bolts. Tighten to specifications if needed and apply white Torque Seal F-900.
	Spring U-Bolts	Tighten all U-bolts to the correct torque.
	Rear Axle Differential	Check oil level and add oil, if necessary.
	Manual Transmission Oil Level	Check oil level and add oil, if necessary.
	All Drain Plugs	Check and tighten all drain plugs and fill plugs.
	Battery Cable(s)	Check all battery cables for correct routing and clamping. Make corrections, if necessary. Tighten connections and reroute cables or add tie-straps as necessary.
	Batteries	Check open circuit voltage for all batteries. If below 12.5 volts, charge to required level.
	Wheel Nuts	Disc-type wheels, tighten all wheel nuts to the correct torque. Spoke-type wheels, check the wheel rim and clamp nuts for the correct torque.
	Exhaust Stack	Check for proper mounting and torque clamps and brackets to specifications
	Aftermarket Treatment Device	Verify that ATD devices are installed, and are operating. Check DEF level.
	Front Hub Level	Check and fill as needed.
	Chassis, Fifth Wheel, Driveshaft	Lube if needed.
	Tire Pressure	Check and adjust as needed.
D. Road Test		
	Parking and Service Brakes	Check for proper operation.
	Clutch Performance	Check for proper free-pedal, clutch disengagement, and for proper clutch brake operation.
	Steering Performance	Check for smooth operation without binding or tires rubbing.
	Speed Control and Throttle Operation	Check throttle response and cruise control operation.
	Two-Speed Axle Operation (if equipped)	Check for proper operation.

Technician Signature: _____   Supervisor: _____   Date: _____

FIGURE 4-11 (continued)

STUDENT SELF-EVALUATION

Check	Level	Competency	Comments
	4	Mastered task	
	3	Competent but need further help	
	2	Needed a lot of help	
	0	Did not understand the task	

INSTRUCTOR EVALUATION

Check	Level	Competency	Comments
	4	Mastered task	
	3	Competent but needs further help	
	2	Requires more training	
	0	Unable to perform task	

JOB SHEET 4.12

Name _____ Station _____ Date _____

Perform a Typical Cooling System Service on a Truck.

Performance Objective(s): Use the checklist that follows to perform a cooling system service on a typical truck. Note that some of the steps may be unnecessary if you are using extended life coolant (ELC) that does not require testing of the coolant chemistry or of the level of antifreeze protection.

ASE Education Foundation Correlation

This job sheet addresses the following ASE Education Foundation task(s):

I. **Diesel Engines**
   A. **General**
      1. Research vehicle service information, including fluid type, vehicle service history, service precautions, and technical service bulletins. (IMMR, TST, MTST) P-1.
   E. **Cooling System**
      1. Check engine coolant type, level, condition, and test coolant for freeze protection and additive package concentration. (IMMR, TST, MTST) P-1.
      2. Verify coolant temperature; check operation of temperature and level sensors, gauge, and/or sending unit. (IMMR) P-1.
      2. Test coolant temperature; test operation of temperature and level sensors, gauge, and/or sending unit; determine needed action. (TST, MTST) P-1.
      3. Inspect and reinstall/replace pulleys, tensioners, and drive belts; adjust drive belts and check alignment. (IMMR, TST, MTST) P-1.
      4. Recover coolant, flush, and refill with recommended coolant/additive package; bleed cooling system. (IMMR, TST, MTST) P-1.
      5. Inspect coolant conditioner/filter assembly for leaks; inspect valves, lines, and fittings; replace as needed. (IMMR, TST, MTST) P-1.
      6. Inspect water pump, hoses, and clamps; determine needed action. (IMMR, TST, MTST) P-1.
      7. Inspect and pressure test cooling system(s); pressure test cap, tank(s), and recovery systems; inspect radiator and mountings; determine needed action. (IMMR, TST, MTST) P-1.
      8. Inspect thermostatic cooling fan system (hydraulic, pneumatic, and electronic) and fan shroud; determine needed action. (IMMR, TST, MTST) P-1.
   F. **Air Induction and Exhaust Systems**
      2. Check air induction system including: cooler assembly, piping, hoses, clamps, and mountings; replace air filter as needed; reset restriction indicator (if applicable). (IMMR, TST) P-1.
      2. Diagnose air induction system problems; inspect, clean, and/or replace cooler assembly, piping, hoses, clamps, and mountings; replace air filter as needed; reset restriction indicator (if applicable). (MTST) P-1.

VI. **Heating, Ventilation, and Air Conditioning (HVAC)**
   D. **Operating Systems and Related Controls**
      1. Verify HVAC system blower motor operation; confirm proper air distribution; confirm proper temperature control; determine needed action. (IMMR, TST, MTST) P-1.

Tools and Materials: A highway truck or tractor unit in running condition, standard personal and shop tools, the following PM procedure, and a pen or pencil

Protective Clothing: Standard shop apparel including coveralls or shop coat, safety glasses, and safety footwear

PROCEDURE
Ensure that the shop LOTO is observed.

1. Complete the following form, noting whether each task has actually been performed or reviewed.

Task completed _____

Cooling System Service

Customer Name _____ RO # _____ VIN # _____ Date _____

Make _____ Model _____ Odometer _____ km/mi

Flush and Fill or Service Only *(circle one)*

Extended Life or Regular Coolant *(circle one)*

Inspection Code: ☑ = OK   A = Adjusted   R = Repaired   X = Replaced   N = Needs Follow-Up

N	Item #	Flush and Fill	Code	Tech #
	1	Drain and refill system with specified antifreeze.		
		Service Only		
	2	Pressure test cooling system.		
	3	Visually inspect engine compartment for fluid leaks (fuel, oil, and water).		
	4	Check hoses and clamps.		
	5	Externally inspect water pump.		
	6	Check coolant filter for leaks. Replace with correct charge filter if unit requires SCA service.		
		Test and **record** coolant protection level and SCA level. Service as necessary.		
	7	Protection _____ SCA units _____		
	8	**Extended Life Coolant: *Add extender if required (480,000 km, 300,000 miles, 6,000 hours).**		
	9	Check all belts, pulleys, and tensioners for wear and adjustment.		
	10	Check radiator condition. Inspect for obstructions, dirt, debris, etc.		
	11	Check fan and shroud for damage.		
	12	Check thermostat operation.		
	13	Check engine fan operation.		

Record of Deficiencies:

Code	Tech.	Date Repaired	#	Service Procedure

Inspected by: _____   Supervisor: _____   Next Service Due: _____

FIGURE 4–12 Cooling system service

STUDENT SELF-EVALUATION

Check	Level	Competency	Comments
	4	Mastered task	
	3	Competent but need further help	
	2	Needed a lot of help	
	0	Did not understand the task	

INSTRUCTOR EVALUATION

Check	Level	Competency	Comments
	4	Mastered task	
	3	Competent but needs further help	
	2	Requires more training	
	0	Unable to perform task	

JOB SHEET 4.13

Name _____ Station _____ Date _____

Perform a Fifth Wheel Service on a Highway Tractor.

Performance Objective(s): Perform a fifth wheel service on a highway tractor. A key to performing this correctly is to properly clean the fifth wheel so that it can be visually inspected.

ASE Education Foundation Correlation

This job sheet addresses the following ASE Education Foundation task(s):

IV Suspension and Steering
  H. Frame and Coupling Devices
  1. Inspect, service, and/or adjust fifth wheel, pivot pins, bushings, locking mechanisms, mounting hardware, air lines, and fittings. (IMMR, TST, MTST) P-1.
  2. Inspect frame and frame members for cracks, breaks, corrosion, distortion, elongated holes, looseness, and damage; determine needed action. (IMMR, TST, MTST) P-1.
  3. Inspect frame hangers, brackets, and cross members. (IMMR) P-3.
  3. Inspect and install frame hangers, brackets, and cross members; determine needed action. (TST, MTST) P-3.
  4. Check pintle hook and mounting (if applicable). (IMMR) P-1.
  4. Inspect, repair, or replace pintle hooks and draw bars (if applicable). (TST, MTST) P-1.

Tools and Materials: A highway truck or tractor unit in running condition, standard personal and shop tools, the following PM procedure, and a pen or pencil

Protective Clothing: Standard shop apparel including coveralls or shop coat, safety glasses, and safety footwear

PROCEDURE
Ensure that the shop LOTO is observed.

1. Complete the following form, noting whether each task has actually been performed or reviewed.

   Task completed _____

Fifth Wheel Service

Customer Name _____ RO # _____ VIN # _____ Date _____

Make _____ Model _____ Odometer _____ km/mi

Fifth Wheel Make _____ Fifth Wheel Model _____

Rebuild or Service (circle one)

Inspection Code: ☑ = OK A = Adjusted R = Repaired X = Replaced N = Needs Follow-Up

N	Item #	Operation Description	Code	Tech #
	1	Steam clean or pressure wash fifth wheel.		
	2	If rebuilding, *remove fifth wheel assembly and follow rebuild instructions as per manufacturer.*		
	3	Thoroughly clean the locking mechanism.		
	4	Inspect for bent, broken, or missing parts.		
	5	Inspect for cracks in the fifth wheel assembly, mounting brackets, and moving parts.		
	6	Check fifth wheel mounting for loose or missing nuts and bolts.		
	7	Check wear limits of pins and bushings as per manufacturer's specifications.		
	8	Lubricate as per fifth wheel manufacturer's recommendations.		
	9	Test the operation of the fifth wheel locking mechanism.		
	10	Check the adjustment of the fifth wheel locks and adjust as per manufacturer's specifications.		
	11	Perform any additional service particular to fifth wheel make and model.		
		Additional Operations		

Record of Deficiencies:

Code	Tech.	Date Repaired	#	Service Procedure

Inspected by: _____ Supervisor: _____ Next Service Due: _____

FIGURE 4–13 Fifth wheel service

STUDENT SELF-EVALUATION

Check	Level	Competency	Comments
	4	Mastered task	
	3	Competent but need further help	
	2	Needed a lot of help	
	0	Did not understand the task	

INSTRUCTOR EVALUATION

Check	Level	Competency	Comments
	4	Mastered task	
	3	Competent but needs further help	
	2	Requires more training	
	0	Unable to perform task	

JOB SHEET 4.14

Name _____ Station _____ Date _____

Perform an A/C Performance Inspection on a Truck.

Performance Objective(s): Use the A/C performance checklist to assess the performance of a truck A/C circuit. You should note that in performing this exercise there are strictly enforced environmental requirements for handling vehicle A/C systems. These fall under both federal and local jurisdictions and you should be in compliance.

ASE Education Foundation Correlation

This job sheet addresses the following ASE Education Foundation task(s):

VI. Heating, Ventilation, and Air Conditioning (HVAC)
  A. General
   1. Research vehicle service information, including refrigerant/oil type, vehicle service history, service precautions, and technical service bulletins. (IMMR, TST, MTST) P-1.
   2. Identify heating, ventilation, and air-conditioning (HVAC) components and configuration. (IMMR, TST, MTST) P-1.
   3. Use appropriate electronic service tool(s) and procedures to check, record, and clear diagnostic codes; interpret digital multimeter (DMM) readings. (IMMR) P-1.
   3. Use appropriate electronic service tool(s) and procedures to diagnose problems; check, record, and clear diagnostic codes; interpret digital multimeter (DMM) readings. (TST, MTST) P-1.
  B. Refrigeration System Components
   1. Inspect A/C compressor drive belts, pulleys, and tensioners; verify proper belt alignment. (IMMR) P-1.
   1. Inspect, remove, and replace A/C compressor drive belts, pulleys, and tensioners; verify proper belt alignment. (TST, MTST) P-1.
   2. Check A/C system operation including system pressures; visually inspect A/C components for signs of leaks; check A/C monitoring system (if applicable). (IMMR, TST, MTST) P-1.
   3. Inspect A/C condenser for airflow restrictions; determine needed action. (IMMR, TST, MTST) P-1.
  C. Heating, Ventilation, and Engine Cooling Systems
   1. Inspect engine cooling system and heater system hoses and pipes; determine needed action. (IMMR, TST, MTST) P-1.
   2. Inspect HVAC system-heater ducts, doors, hoses, cabin filters, and outlets; determine needed action. (IMMR, TST, MTST) P-1.
   3. Identify the source of A/C system odors; determine needed action. (IMMR, TST, MTST) P-1.
  D. Operating Systems and Related Controls
   1. Verify blower motor operation; confirm proper air distribution; confirm proper temperature control; determine needed action. (IMMR, TST, MTST) P-1.

Tools and Materials: A highway truck or tractor unit in running condition, standard personal and shop tools, the following PM procedure, and a pen or pencil

Protective Clothing: Standard shop apparel including coveralls or shop coat, safety glasses, and safety footwear

PROCEDURE
Ensure that the shop LOTO is observed.

1. Complete the following form, noting whether each task has actually been performed or reviewed.

Task completed _____

Air-Conditioning Performance Service

Customer Name _____ RO # _____ VIN # _____ Date _____

Make _____ Model _____ Odometer _____ km/mi

Inspection Code: ✓ = OK   A = Adjusted   R = Repaired   X = Replaced   N = Needs Follow-Up

N	Item #	In Cab Inspection	Code	Tech #
	1	Inspect evaporator core and heater core, if combined. Clean away any debris. Verify drain is clear.		
	2	Check radiator, CAC and AC condenser condition, mounting, obstructions, dirt, debris, etc.		
	3	Check fan and shroud for damage.		
	4	Check all belts, pulleys, and tensioners for wear and adjustment.		
	5	Inspect hose condition; check routing and clipping.		
	6	Inspect AC wiring harness and connections.		
	7	Inspect compressor. Check for loose or broken wires. Check mounting bracket and hardware.		
	8	Check heater valves (be sure heater valves are closing fully) and controls for proper operation.		
	9	Connect AC manifold gauges.		
	10	Start vehicle and operate the air-conditioning system with engine at 1,200–1,500 RPM.		
	11	Check receiver/dryer service indicator (Note: should be serviced every 2 years).		
	12	Check high- and low-pressure gauge readings to ensure proper operation.		
	13	Record running system pressure: High side _____ Low side _____		
	14	If system requires service, refer to RP 413 (R-12) or RP 418 (R-134a).		
	15	Record running system pressure after service is performed: High side _____ Low side _____		
	16	Verify proper clutch operation. Check for any unusual noises.		
	17	Check for proper blower speed control and measure air temperature at vent closest to the evaporator.		
	18	Record duct outlet temperature _____		

Record of Deficiencies:

Code	Tech.	Date Repaired	#	Service Procedure

Inspected by: _____   Supervisor: _____   Next Service Due: _____

FIGURE 4–14 A/C performance service

STUDENT SELF-EVALUATION

Check	Level	Competency	Comments
	4	Mastered task	
	3	Competent but need further help	
	2	Needed a lot of help	
	0	Did not understand the task	

INSTRUCTOR EVALUATION

Check	Level	Competency	Comments
	4	Mastered task	
	3	Competent but needs further help	
	2	Requires more training	
	0	Unable to perform task	

ONLINE TASKS

Use a search engine to access the following Web pages and describe the role each plays in the trucking industry:

1. OSHA
2. NHTSA
3. CVSA

Answer These Questions:

1. What role does each organization play in making the workplace or our roads safer?
2. What is an FMVSS?
3. What is the purpose of a TMC RP paper?

STUDY TIPS

Identify five key points in Chapter 4. Try to be as brief as possible.

Key point 1 _____

Key point 2 _____

Key point 3 _____

Key point 4 _____

Key point 5 _____

5 Fundamentals of Electricity

Objectives

After reading this chapter, you should be able to:
- Define the terms *electricity* and *electronics*.
- Describe atomic structure.
- Outline how some of the chemical and electrical properties of atoms are defined by the number of electrons in their outer shells.
- Outline the properties of conductors, insulators, and semiconductors.
- Describe the characteristics of *static electricity*.
- Define what is meant by the *conventional* and *electron theories* of current flow.
- Describe the characteristics of magnetism and the relationship between electricity and magnetism.
- Describe how electromagnetic field strength is measured in common electromagnetic devices.
- Define what is meant by an electrical circuit and the terms *voltage*, *resistance*, and *current flow*.
- Outline the components required to construct a typical electrical circuit.
- Perform electrical circuit calculations using Ohm's Law.
- Identify the characteristics of DC and AC.
- Describe some methods of generating a current flow in an electrical circuit.
- Describe and apply Kirchhoff's First and Second Laws.

PRACTICE QUESTIONS

1. Which of the following is true of an electron?
    a. It has a positive charge.
    b. It has a negative charge.
    c. It is part of the nucleus of an atom.
    d. It has more mass than a proton.

2. The conventional theory of electricity says that current flows from:
    a. positive to negative
    b. negative to positive
    c. low potential to high potential
    d. ground to a circuit load

3. Which of the following would have the least ability to conduct electricity?
    a. gold
    b. copper
    c. aluminum
    d. rubber

4. Which theory of electrical current flow states that current flows from a negative location to a positive location in an electrical circuit?
    a. Franklin's Law
    b. Electron Theory
    c. Conventional Theory
    d. Kirchhoff's Law

5. Who was responsible for building the world's first electrical power station and distribution grid in New York City?
   a. Edison
   b. Ford
   c. Faraday
   d. Gilbert
6. Match the words in the right column to those on the left:

pressure	watts
flow	ohms
resistance	volts
power	amps

7. What type of signal is produced by an inductive pulse generator?
   a. sawtooth
   b. square wave
   c. AC
   d. DC
8. How should an ohmmeter be connected into an electrical circuit?
   a. in series, circuit energized
   b. in parallel, circuit energized
   c. in series, circuit open
   d. in parallel, circuit open
9. When a simple series circuit is converted into a parallel circuit by adding another circuit path and load, what must happen to total circuit resistance?
   a. It increases.
   b. It decreases.
   c. It is not affected.
10. Which of the following determines the electromagnetic field strength developed in a coil?
    a. amperage
    b. number of windings
    c. both a and b
    d. neither a nor b

JOB SHEET 5.1

Name _____ Station _____ Date _____

Ohm's and Watt's Laws Familiarization.

Performance Objective(s): Measure circuit values at random in a truck electrical circuit using a DMM. Use Ohm's and Watt's Laws to check the circuit values and calculate power requirements.

ASE Education Foundation Correlation
This job sheet addresses the following ASE Education Foundation task(s):

V. Electrical/Electronic Systems
  A. General
  1. Research vehicle service information, including vehicle service history, service precautions, and technical service bulletins. (IMMR, TST, MTST) P-1.
  2. Demonstrate knowledge of electrical/electronic series, parallel, and series-parallel circuits using principles of electricity (Ohm's Law). (IMMR, TST, MTST) P-1.
  3. Demonstrate proper use of test equipment when measuring source voltage, voltage drop (including grounds), current flow, continuity, and resistance. (IMMR, TST, MTST) P-1.
  4. Demonstrate knowledge of the causes and effects of shorts, grounds, opens, and resistance problems in electrical/electronic circuits. (IMMR) P-1.
  4. Demonstrate knowledge of the causes and effects of shorts, grounds, opens, and resistance problems in electrical/electronic circuits; identify and locate faults in electrical/electronic circuits. (TST, MTST) P-1.
  5. Use wiring diagrams to trace electrical/electronic circuits. (IMMR) P-1.
  5. Use wiring diagrams during the diagnosis (troubleshooting) of electrical/electronic circuit problems. (TST, MTST) P-1.
  11. Identify electrical/electronic system components and configuration. (IMMR, TST, MTST) P-1.

Tools and Materials: A truck with a functional electrical circuit or a simple breadboard circuit, a DMM, shop tools, and a calculator

Protective Clothing: Standard shop apparel including coveralls or shop coat, safety glasses, and safety footwear

PROCEDURE

Reference Figure 5–1 and then fill in the data for two of each of the three fields listed below. Calculate the third using the formulae in Figure 5–1. Check your result with a DMM.

OHM'S LAW CALCULATION

$A = \dfrac{V}{R}$

$V = A \cdot R$

$R = \dfrac{V}{A}$

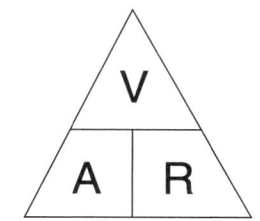

Points to remember: If voltage (V) is constant and resistance (R) increases, the amperage (A) will decrease. If voltage (V) is constant and resistance (R) decreases, amperage (A) will increase.

FIGURE 5–1 Ohm's and Watt's Laws

WATT'S LAW (POWER) CALCULATION

$W = V \cdot A$

$V = \dfrac{W}{A}$

$A = \dfrac{W}{V}$

$\left.\begin{matrix} \\ \\ \\ \end{matrix}\right\}$

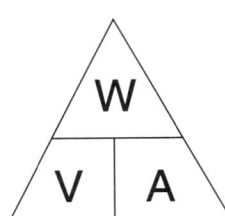

FIGURE 5–1 (Continued)

Circuit exercises:

1. A_____

   V_____

   R_____

2. A_____

   V_____

   R_____

3. A_____

   V_____

   R_____

Power calculations:

1. A_____

   V_____

   W_____

2. A_____

   V_____

   W_____

3. A_____

   V_____

   W_____

Task completed _____

JOB SHEET 5.2

Name _____ Station _____ Date _____

Perform Voltage and Resistance Measurements on a Truck Chassis Electrical Circuit.

Performance Objective(s): Use a DMM to perform voltage and resistance measurements on a truck chassis electrical circuit. Use the DMM readings to calculate and confirm the application of Ohm's Law.

ASE Education Foundation Correlation

This job sheet addresses the following ASE Education Foundation task(s):

V. Electrical/Electronic Systems
  A. General
  1. Research vehicle service information, including vehicle service history, service precautions, and technical service bulletins. (IMMR, TST, MTST) P-1.
  2. Demonstrate knowledge of electrical/electronic series, parallel, and series-parallel circuits using principles of electricity (Ohm's Law). (IMMR, TST, MTST) P-1.
  3. Demonstrate proper use of test equipment when measuring source voltage, voltage drop (including grounds), current flow, continuity, and resistance. (IMMR, TST, MTST) P-1.
  4. Demonstrate knowledge of the causes and effects of shorts, grounds, opens, and resistance problems in electrical/electronic circuits. (IMMR) P-1.
  4. Demonstrate knowledge of the causes and effects of shorts, grounds, opens, and resistance problems in electrical/electronic circuits; identify and locate faults in electrical/electronic circuits. (TST, MTST) P-1.
  5. Use wiring diagrams to trace electrical/electronic circuits. (IMMR) P-1.
  5. Use wiring diagrams during the diagnosis (troubleshooting) of electrical/electronic circuit problems. (TST, MTST) P-1.
  6. Measure parasitic (key-off) battery drain. (IMMR) P-1.
  6. Measure parasitic (key-off) battery drain; determine needed action. (TST, MTST) P-1.
  7. Demonstrate knowledge of the function, operation, and testing of fusible links, circuit breakers, relays, solenoids, diodes, and fuses. (IMMR) P-1.
  7. Demonstrate knowledge of the function, operation, and testing of fusible links, circuit breakers, relays, solenoids, diodes, and fuses; perform inspection and testing; determine needed action. (TST, MTST) P-1.
  8. Inspect, repair (including solder repair), and/or replace connectors, seals, terminal ends, and wiring; verify proper routing and securement. (IMMR) P-1.
  8. Inspect, test, repair (including solder repair), and/or replace components, connectors, seals, terminal ends, harnesses, and wiring; verify proper routing and securement; determine needed action. (TST, MTST) P-1.
  9. Use appropriate electronic service tool(s) and procedures to check, record, and clear diagnostic codes; interpret digital multimeter (DMM) readings. (IMMR) P-2.
  9. Use appropriate electronic service tool(s) and procedures to diagnose problems; check, record, and clear diagnostic codes; interpret digital multimeter (DMM) readings. (TST, MTST) P-2.

  B. Battery System
  1. Identify battery type and system configuration. (IMMR, TST, MTST) P-1.
  2. Confirm proper battery capacity for application; perform battery state-of-charge test; perform battery capacity test, determine needed action. (IMMR, TST, MTST) P-1.
  3. Inspect battery, battery cables, connectors, battery boxes, mounts, and hold-downs; determine needed action. (IMMR, TST, MTST) P-1.

  C. Starting System
  1. Demonstrate understanding of starter system operation. (IMMR, TST, MTST) P-1.
  2. Perform starter circuit cranking voltage and voltage drop tests. (IMMR) P-1.
  2. Perform starter circuit cranking voltage and voltage drop tests; determine needed action. (TST, MTST) P-1.

3. Inspect starter control circuit switches, relays, connectors, terminals, wires, and harnesses (including over-crank protection). (IMMR) P-1.
3. Inspect starter control circuit switches, relays, connectors, terminals, wires, and harnesses (including over-crank protection); determine needed action. (TST, MTST) P-1.

D. Charging System
1. Identify and understand operation of the generator (alternator). (IMMR, TST, MTST) P-1.
2. Check instrument panel mounted voltmeters and/or indicator lamps. (IMMR) P-1.
2. Check instrument panel mounted voltmeters and/or indicator lamps; determine needed action. (TST, MTST) P-1.
4. Inspect cables, wires, and connectors in the charging circuit. (IMMR) P-1.
4. Inspect cables, wires, and connectors in the charging circuit; determine needed action. (TST, MTST) P-1.
5. Perform charging system voltage and amperage output tests; perform AC ripple test. (IMMR) P-1.
5. Perform charging system voltage and amperage output tests; perform AC ripple test; determine needed action. (TST, MTST) P-1.

Tools and Materials: A truck with a functional electrical circuit, a DMM, shop tools, and a calculator

Protective Clothing: Standard shop apparel including coveralls or shop coat, safety glasses, and safety footwear

PROCEDURE
Ensure that the shop LOTO is observed.

1. Record the following vehicle data:

   OEM _____ Model year _____ Engine _____ VIN _____

   Task completed _____

2. Select the Volts DC mode on the DMM. If the DMM is not auto-ranging, set it on the 18V scale. Connect the DMM across the battery terminals and record the voltage value.

   Battery voltage_____

   Task completed _____

3. With the engine off, turn on all the dash-controlled electrical loads on the truck and once again record the battery voltage. Turn the electrical loads off.

   Battery voltage_____

   Task completed _____

4. Access one of the truck clearance lights so the terminals on the pigtail can be probed. Do not perform this task if sealed terminal LEDs are used. Turn the clearance lights on. Ground one DMM probe. Now use the other DMM probe to check the voltage, first on the positive side of the pigtail, then on the negative, recording the values as V1 and V2.

   V1_____ V2_____

   Sealed LED light circuit _____

   Task completed _____

5. Switch off the clearance lights and disconnect the vehicle batteries either by using the isolator switch or by removing the battery terminals.

   Task completed _____

6. Measure the resistance using the DMM through the clearance light and record the value. Set the DMM into resistance test mode and place one probe on the positive terminal and the other on the negative.

   Use Ohm's Law and the first battery voltage value you measured (at the beginning of this job sheet) to calculate current flow through the light. Use a calculator. Record the values.

   Clearance light resistance_____ Battery voltage _____ Amperage _____

   Task completed _____

7. With the batteries still disconnected, measure the resistance values through some typical loads on the truck electrical system. Use those electrical loads that are easiest to access on the truck and take care not to damage any sealed connectors. Record the resistance values.

   Starter motor_____ Heater motor_____ Headlight_____

   Block heater_____ Fuel sending unit_____

   Using Ohm's Law, the battery voltage you measured in the first test, and a calculator, perform some amperage calculations.

   Task completed _____

8. Reconnect the truck batteries and verify the operation of any components you have had disconnected from the circuit.

   Task completed _____

Review Task: CIRCUIT TEST EQUIPMENT CONNECTIONS

Study Figure 5–2 and memorize it. Knowing how to make these connections in electrical circuits forms the basis of most electrical test circuit testing. Put this knowledge into practice using some truck electrical circuits.

Fundamentals of Electricity

CIRCUIT TEST EQUIPMENT CONNECTIONS

Voltage Measurements

When checking voltage in an electrical system component or circuit, connect the voltmeter in parallel to the system component or circuit.

Common values used when measuring voltage are kV (kilo-volts, usually used for spark plug and ignition circuit voltages), V (volts, usually used for battery, vehicle lighting, and electrical motor voltages) and mV (milli-volts, used for some electronic inputs/sensor input voltages).

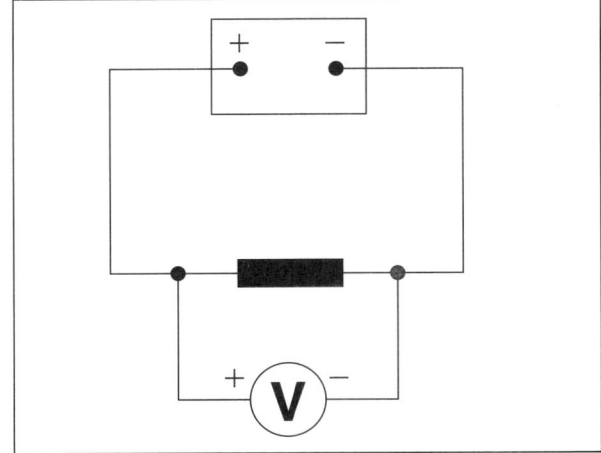

Amperage Measurements

When checking amperage in an electrical system component or circuit, connect the ammeter in series with the system component or circuit. Connect the positive terminal on the ammeter to the point with the highest potential. If the amperage is high in the circuit, use an inductive type ≠ammeter to perform checks. Common values used when measuring amperage are A (amps, usually used for battery, lighting, and electrical motor amperages), mA (milli-amp, usually used for electronic power supplies and signals) and µA (micro-amp, usually used for electronic amplifier inputs such as temperature sensors).

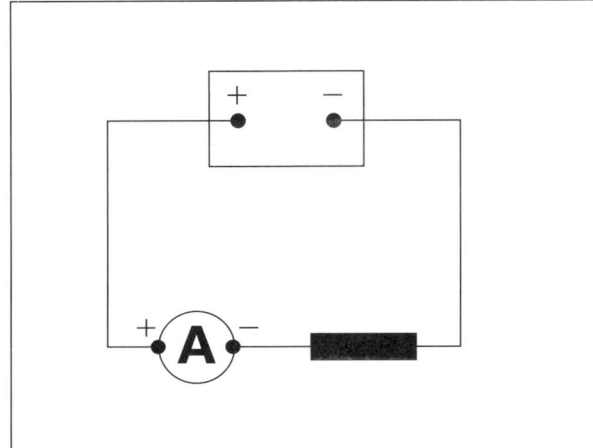

Resistance Measurements

When performing an electrical system component resistance check, connect the ohmmeter in series with the component to be tested with no potential to the circuit.

When performing a circuit continuity check, and if all the electrical connectors have been checked and are satisfactory, the ohmmeter should indicate close to zero (0) ohms. The ohmmeter will show infinity if the circuit is open.

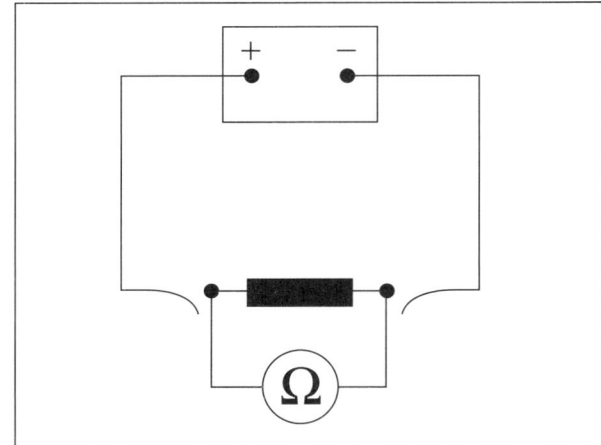

FIGURE 5-2 Circuit test equipment connections

STUDENT SELF-EVALUATION

Check	Level	Competency	Comments
	4	Mastered task	
	3	Competent but need further help	
	2	Needed a lot of help	
	0	Did not understand the task	

INSTRUCTOR EVALUATION

Check	Level	Competency	Comments
	4	Mastered task	
	3	Competent but needs further help	
	2	Requires more training	
	0	Unable to perform task	

ONLINE TASKS

Locate the "howstuffworks" Web site. Record this URL on your Favorites/Bookmark file. Now get the "howstuffworks" explanation of how the following devices function:

1. Light bulb
2. Electric motor
3. Battery

STUDY TIPS

Identify 5 key points in Chapter 5. Try to be as brief as possible.

Key point 1 _____

Key point 2 _____

Key point 3 _____

Key point 4 _____

Key point 5 _____

Fundamentals of Electricity

6 Fundamentals of Electronics and Computers

Objectives

After reading this chapter, you should be able to:

- Outline some of the developmental history of electronics.
- Describe how an electrical signal can be used to transmit information.
- Define the term *pulse width modulation*.
- Define the principle of operation of N- and P-type semiconductors.
- Outline the operating principles and applications of diodes.
- Describe the construction and operation of a typical transistor.
- Describe what is meant by the *optical spectrum*.
- Identify some commonly used optical components used in electronic circuitry.
- Explain what is meant by an *integrated circuit* and outline its application in on-board vehicle electronics.
- Define the role of gates in electronic circuits.
- Describe the operating modes of some common gates used in electrical circuits, including AND, OR, and NOT gates.
- Interpret a *truth table* that defines the outcomes of gates in an electronic circuit.
- Explain why the binary numeric system is used in computer electronics.
- Define the role of an electronic control module in an electronic management system.
- Outline the distinct stages of a computer processing cycle.
- Describe the data retention media used in vehicle ECUs.
- Demonstrate an understanding of input circuits on a vehicle electronic system.
- Troubleshoot a potentiometer-type TPS.
- Describe the operating principles of some accident avoidance systems.

PRACTICE QUESTIONS

1. Which of the following is not a solid-state component?
   a. transistor
   b. diode
   c. solenoid
   d. zener diode

2. An electrical signal that varies according to percentage of ON and OFF time would be described as:
   a. pulse width modulated
   b. variable frequency
   c. sinusoidal
   d. alternating current

3. Which of the following are semiconductors?
   a. germanium
   b. gold
   c. silicon
   d. platinum

4. Before a semiconductor can be used in a solid-state device, it must first be:
   a. purified
   b. doped
   c. decomposed
   d. activated

5. Which type of diode produces an electrical response when subjected to light?
   a. zener diode
   b. LED
   c. small signal diode
   d. photodiode

6. What type of transistor would be used to switch diesel electronic unit injectors?
   a. power transistor
   b. FET
   c. SCR
   d. MOSFET

7. Which DMM test mode would be required to bench test transistors?
   a. V-DC
   b. V-AC
   c. ohms
   d. milliamps

8. What would be required of a two-switch AND gate to produce an output?
   a. both switches off
   b. both switches on
   c. either switch on

9. Which of the following devices is often used to measure temperature?
   a. rheostat
   b. potentiometer
   c. pulse generator
   d. thermistor

10. Which type of thermistor would be most likely used in a truck input circuit?
    a. NTC
    b. PTC
    c. AMP
    d. CRT

11. Express the following byte data in a different format: 0000 1000.
    a. 4
    b. 8
    c. 256
    d. 1,000

12. Which of the following voltage values would be most commonly used as V-Ref?
    a. 1 V
    b. 5 V
    c. 12.6 V
    d. ±100 V

13. Which of the following input sensors would be used to signal pressure values?
    a. piezo-resistive
    b. pulse generator
    c. thermistor
    d. potentiometer

14. How does Doppler effect determine relative velocity?
    a. by bouncing radiated return signals
    b. by analyzing frequency shift on return signals
    c. noise analysis
    d. light spectrum analysis
15. Which of the following terms is the truck technology term used to describe what happens when two or more chassis computers "talk" to each other?
    a. networking
    b. multiplexing
    c. chatting
    d. texting

JOB SHEET 6.1

Name _____ Station _____ Date _____

Construct Some Simple Gate Circuits on a Breadboard.

Performance Objective(s): Understand the building blocks of electronic signal routing by building some simple gate circuits on a breadboard (circuit board) using a power supply and simple components.

ASE Education Foundation Correlation

This job sheet addresses the following ASE Education Foundation task(s):

V. Electrical/Electronic Systems
  A. General
  2. Demonstrate knowledge of electrical/electronic series, parallel, and series-parallel circuits using principles of electricity (Ohm's Law). (IMMR, TST, MTST) P-1.
  3. Demonstrate proper use of test equipment when measuring source voltage, voltage drop (including grounds), current flow, continuity, and resistance. (IMMR, TST, MTST) P-1.
  4. Demonstrate knowledge of the causes and effects of shorts, grounds, opens, and resistance problems in electrical/electronic circuits. (IMMR) P-1.
  4. Demonstrate knowledge of the causes and effects of shorts, grounds, opens, and resistance problems in electrical/electronic circuits; identify and locate faults in electrical/electronic circuits. (TST, MTST) P-1.
  5. Use wiring diagrams to trace electrical/electronic circuits. (IMMR) P-1.
  5. Use wiring diagrams during the diagnosis (troubleshooting) of electrical/electronic circuit problems. (TST, MTST) P-1.
  9. Use appropriate electronic service tool(s) and procedures to check, record, and clear diagnostic codes; interpret digital multimeter (DMM) readings. (IMMR) P-2.
  9. Use appropriate electronic service tool(s) and procedures to diagnose problems; check, record, and clear diagnostic codes; interpret digital multimeter (DMM) readings. (TST, MTST) P-2.
  11. Identify electrical/electronic system components and configuration. (IMMR, TST, MTST) P-1.

  F. Instrument Cluster and Driver Information Systems
  2. Identify the sensor/sending units, gauges, switches, relays, bulbs/LEDs, wires, terminals, connectors, sockets, printed circuits, and control components/modules of the instrument cluster, driver information system, and warning systems. (IMMR) P-2.
  2. Identify faults in the sensor/sending units, gauges, switches, relays, bulbs/LEDs, wires, terminals, connectors, sockets, printed circuits, and control components/modules of the instrument cluster, driver information systems, and warning systems; determine needed action. (TST, MTST) P-1.

- Narrative for science-related academic skills

Tools and Materials: Small circuit board, normally open (NO) and normally closed (NC) push-button switches, power supply, a light bulb, and interconnecting wiring.

Protective Clothing: None required if performed in a classroom setting; standard shop apparel coveralls or shop coat, safety glasses, and safety footwear.

PROCEDURE

To build the following circuits, reference the figure preceding each procedure.

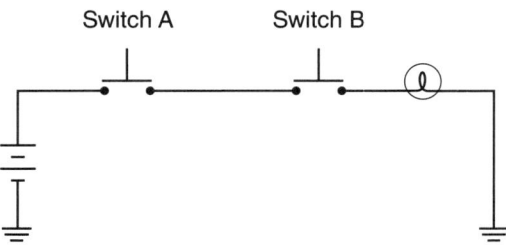

FIGURE 6–1 AND gate.

1. Construct an AND gate. Reference Figure 6–1. Obtain a power supply unit. This can be a battery or transformer, but keep the voltage values below 12 V. This series circuit requires two NO switches and a light bulb arranged as shown in the figure. Note that switch A and switch B must be closed before the light bulb will illuminate.

   Task completed _____

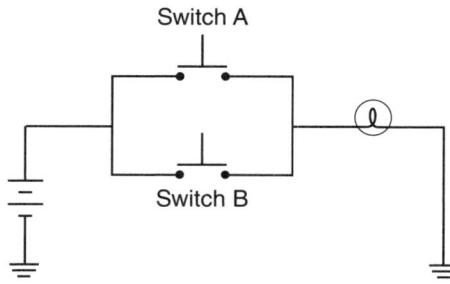

FIGURE 6–2 OR gate.

2. Construct an OR gate. Reference Figure 6–2. Build a circuit as shown with the parallel arrangement of the two NO switches. Note that closing either switch results in illuminating the light bulb.

   Task completed _____

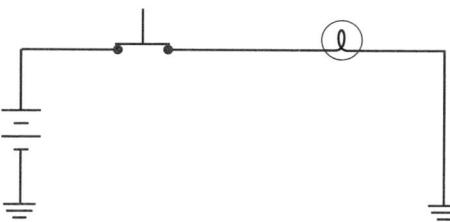

FIGURE 6–3 NOT gate.

3. Construct a NOT gate. Reference Figure 6–3. Build a series circuit as shown with a single NC gate. How does this differ from the previous two exercises?

   Task completed _____

STUDENT SELF-EVALUATION

Check	Level	Competency	Comments
	4	Mastered task	
	3	Competent but need further help	
	2	Needed a lot of help	
	0	Did not understand the task	

INSTRUCTOR EVALUATION

Check	Level	Competency	Comments
	4	Mastered task	
	3	Competent but needs further help	
	2	Requires more training	
	0	Unable to perform task	

JOB SHEET 6.2

Name _____ Station _____ Date _____

Bench Test Electronic Circuit Components.

Performance Objective(s): Perform some resistance tests on some components used in truck electronic circuits.

ASE Education Foundation Correlation
This job sheet addresses the following ASE Education Foundation task(s):

V. Electrical/Electronic Systems
  A. General
  2. Demonstrate knowledge of electrical/electronic series, parallel, and series-parallel circuits using principles of electricity (Ohm's Law). (IMMR, TST, MTST) P-1.
  3. Demonstrate proper use of test equipment when measuring source voltage, voltage drop (including grounds), current flow, continuity, and resistance. (IMMR, TST, MTST) P-1.
  4. Demonstrate knowledge of the causes and effects of shorts, grounds, opens, and resistance problems in electrical/electronic circuits. (IMMR) P-1.
  4. Demonstrate knowledge of the causes and effects of shorts, grounds, opens, and resistance problems in electrical/electronic circuits; identify and locate faults in electrical/electronic circuits. (TST, MTST) P-1.
  5. Use wiring diagrams to trace electrical/electronic circuits. (IMMR) P-1.
  5. Use wiring diagrams during the diagnosis (troubleshooting) of electrical/electronic circuit problems. (TST, MTST) P-1.
  9. Use appropriate electronic service tool(s) and procedures to check, record, and clear diagnostic codes; interpret digital multimeter (DMM) readings. (IMMR) P-2.
  9. Use appropriate electronic service tool(s) and procedures to diagnose problems; check, record, and clear diagnostic codes; interpret digital multimeter (DMM) readings. (TST, MTST) P-2.
  11. Identify electrical/electronic system components and configuration. (IMMR, TST, MTST) P-1.

Tools and Materials: A DMM, OEM diagnostic literature, hot water, and an assortment of sensors from truck electronic circuits. The components used may be from ABS/ATC, transmissions, or engine systems, and some of them should be functional.

Protective Clothing: Standard shop apparel including coveralls or shop coat, safety glasses, and safety footwear.

PROCEDURE

1. Set the DMM in resistance test mode. Where possible, try to obtain the OEM test data for the components to be tested. Many of the electronic circuit components found in training facilities have already failed, so it is important to have some components that are known to function properly.

2. Use the following chart to record the OEM-specified values (from service manual/PC download) and the test values you measure. Use Figure 6–4 as a reference.

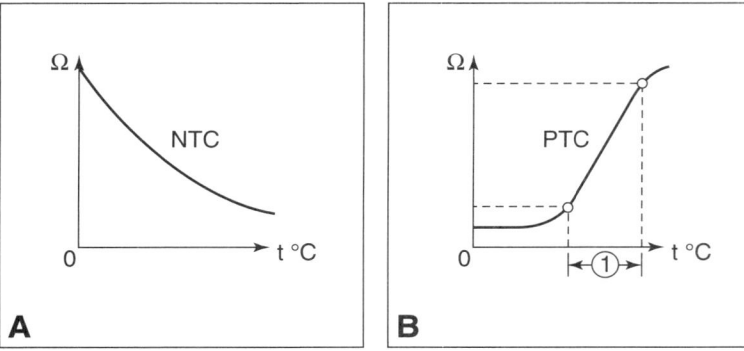

Thermistors change their resistance with temperature: The NTC thermistor (A) will decrease resistance as temperature increases, while the PTC (B) increases resistance with temperature.

FIGURE 6–4 NTC and PTC thermistor performance.

Sensor Component	OEM -R	Test -R
NTC thermistor at room temperature		
NTC thermistor after immersion in hot water		
Pulse generator sensor		
Piezo-resistive pressure sensor		
Variable-capacitance pressure sensor		
Potentiometer		

Task completed _____

Review Task: Check out the operating principle of a diode. Use a DMM and familiarize yourself with the operation of a diode, referencing Figure 6–5.

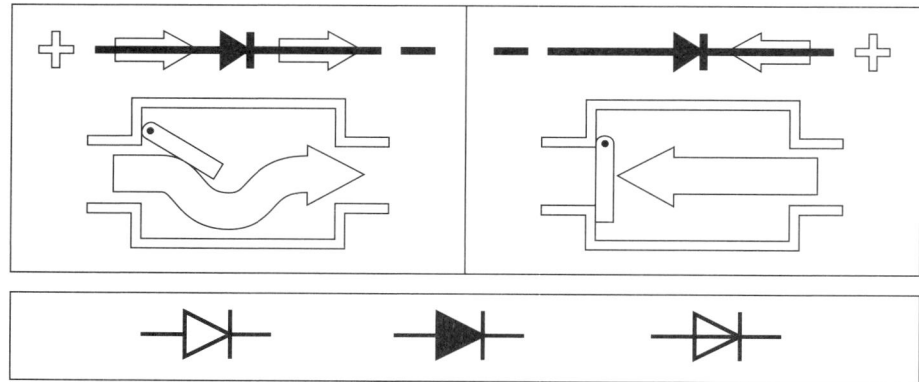

FIGURE 6–5 Diode principle. Diodes only allow current to flow in one direction (forward bias) and block in the other direction (reverse bias). This means that a diode is the electrical equivalent to a one-way check valve in a hydraulic circuit. Diodes are used in converting AC to DC in the alternator rectifier circuit.

STUDENT SELF-EVALUATION

Check	Level	Competency	Comments
	4	Mastered task	
	3	Competent but need further help	
	2	Needed a lot of help	
	0	Did not understand the task	

INSTRUCTOR EVALUATION

Check	Level	Competency	Comments
	4	Mastered task	
	3	Competent but needs further help	
	2	Requires more training	
	0	Unable to perform task	

ONLINE TASKS

Get onto the "howstuffworks" and Wikipedia Web sites. Check out the explanations for the following devices/concepts:

1. Diode
2. Transistor
3. Boolean algebra (see if you can connect this with what we learned in Chapter 6)

STUDY TIPS

Identify five key points in Chapter 6. Try to be as brief as possible.

Key point 1 _____

Key point 2 _____

Key point 3 _____

Key point 4 _____

Key point 5 _____

7 Batteries

Objectives

After reading this chapter, you should be able to:

- Define the role of a battery in a vehicle electrical system.
- Outline the construction of standard, maintenance-free, gelled electrolyte, and absorbed glass mat (AGM) batteries.
- Describe the chemical action within the battery during the charging and discharging cycles.
- Outline how batteries are arranged in multiple battery banks in truck chassis.
- Verify the performance of a lead-acid battery using a voltmeter, hydrometer, refractometer, and carbon pile tester.
- Perform a simple voltage drop test on a battery post to battery clamp.
- Analyze maintenance-free battery condition using an integral hydrometer sight glass.
- Describe the procedure required to charge different types of batteries.
- Jump-start vehicles with dead batteries using another vehicle and generator methods.
- Outline how batteries should be safely stored out of chassis.
- Describe the operating principles of ultracapacitors and how they can be used in hybrid electric vehicles.
- Outline the features of typical vehicle battery management information systems.

PRACTICE QUESTIONS

1. Technician A says that battery gassing produces an explosive mixture of hydrogen and oxygen and that great care should be taken any time a battery is being charged. Technician B says that gassing only occurs during battery discharge cycles on maintenance-free batteries. Who is correct?
    a. Technician A only
    b. Technician B only
    c. both A and B
    d. neither A nor B

2. Rapid charge and discharge cycling of a battery is known as:
    a. transient cycling
    b. deep cycling
    c. sulfation
    d. aeration

3. Which of the following battery electrolyte conditions would be most likely to cause a battery to freeze in winter conditions?
    a. electrolyte specific gravity at 1.265
    b. electrolyte specific gravity at 1.225
    c. a sulfated battery
    d. a 50 percent charged battery

4. Which of the following is true of vehicle batteries?
    a. store electrical energy produced by the alternator
    b. help stabilize system voltage
    c. use a galvanic operating principle
    d. all of the above

5. Which of the following batteries categories would function upside down?
   a. wet cell lead-acid
   b. maintenance-free lead-acid
   c. AGM
   d. No vehicle batteries will function upside down.

6. The positive plates in a lead-acid battery are known as:
   a. anodes
   b. cathodes
   c. polypropylene cases
   d. antimony

7. The negative plates in a lead-acid battery are known as:
   a. anodes
   b. cathodes
   c. polypropylene cases
   d. antimony

8. Which of the following instruments would produce the most accurate measurement of the specific gravity of battery electrolyte?
   a. digital multimeter (DMM)
   b. hydrometer
   c. refractometer
   d. testlight

9. Technician A says that when using a hydrometer to test the specific gravity of battery electrolyte, the reading must be temperature corrected. Technician B says that a refractometer measures the refractive index of electrolyte that increases proportionally with density. Who is correct?
   a. Technician A only
   b. Technician B only
   c. both A and B
   d. neither A nor B

10. Technician A says that most truck OEMs use the CCA specification of a battery as their primary method of rating it. Technician B says that some fleets use ampere-hour rating of a battery when spec'ing it. Who is correct?
    a. Technician A only
    b. Technician B only
    c. both A and B
    d. neither A nor B

JOB SHEET 7.1

Name _____ Station _____ Date _____

Battery Performance Test.

Performance Objective(s): Perform a comprehensive battery test to verify performance effectiveness.

ASE Education Foundation Correlation

This job sheet addresses the following ASE Education Foundation task(s):

V. Electrical/Electronic Systems
   A. General
   1. Research vehicle service information, including vehicle service history, service precautions, and technical service bulletins. (IMMR, TST, MTST) P-1.
   2. Demonstrate knowledge of electrical/electronic series, parallel, and series-parallel circuits using principles of electricity (Ohm's Law). (IMMR, TST, MTST) P-1.
   3. Demonstrate proper use of test equipment when measuring source voltage, voltage drop (including grounds), current flow, continuity, and resistance. (IMMR, TST, MTST) P-1.
   4. Demonstrate knowledge of the causes and effects of shorts, grounds, opens, and resistance problems in electrical/electronic circuits. (IMMR) P-1.
   4. Demonstrate knowledge of the causes and effects of shorts, grounds, opens, and resistance problems in electrical/electronic circuits; identify and locate faults in electrical/electronic circuits. (TST, MTST) P-1.
   5. Use wiring diagrams to trace electrical/electronic circuits. (IMMR) P-1.
   5. Use wiring diagrams during the diagnosis (troubleshooting) of electrical/electronic circuit problems. (TST, MTST) P-1.
   6. Measure parasitic (key-off) battery drain. (IMMR) P-1.
   6. Measure parasitic (key-off) battery drain; determine needed action. (TST, MTST) P-1.
   7. Demonstrate knowledge of the function, operation, and testing of fusible links, circuit breakers, relays, solenoids, diodes, and fuses. (IMMR) P-1.
   7. Demonstrate knowledge of the function, operation, and testing of fusible links, circuit breakers, relays, solenoids, diodes, and fuses; perform inspection and testing; determine needed action. (TST, MTST) P-1.
   9. Use appropriate electronic service tool(s) and procedures to check, record, and clear diagnostic codes; interpret digital multimeter (DMM) readings. (IMMR) P-2.
   9. Use appropriate electronic service tool(s) and procedures to diagnose problems; check, record, and clear diagnostic codes; interpret digital multimeter (DMM) readings. (TST, MTST) P-2.

   B. Battery System
   1. Identify battery type and system configuration. (IMMR, TST, MTST) P-1.
   2. Confirm proper battery capacity for application; perform battery state-of-charge test; perform battery capacity test, determine needed action. (IMMR, TST, MTST) P-1.
   3. Inspect battery, battery cables, connectors, battery boxes, mounts, and hold-downs; determine needed action. (IMMR, TST, MTST) P-1.
   4. Charge battery using appropriate method for battery type. (IMMR, TST, MTST) P-1.
   6. Identify low voltage disconnect (LVD) systems. (IMMR, TST) P-1.
   6. Identify low voltage disconnect (LVD) systems. (MTST) P-1.

Tools and Materials: Any type of heavy-duty truck battery, preferably not of the maintenance-free type; refractometer (or hydrometer), carbon pile load tester, battery charging station, and DMM.

Protective Clothing: Standard shop apparel including coveralls or shop coat, safety glasses, and safety footwear.

PROCEDURE
Ensure that the shop LOTO is observed.

Visual inspection: Note any battery case deformation or damage. Clean battery terminals using wire brush, and clean battery case using mild baking soda solution.

Task completed _____

1. Test electrolyte and record readings:

    Cell #1 _____

    Cell #2 _____

    Cell #3 _____

    Cell #4 _____

    Cell #5 _____

    Cell #6 _____

    Task completed _____

2. Load test, recording data:

    Battery rating _____ CCA

    Load applied _____ amps

    Battery voltage before load test _____ VDC

    Battery voltage at completion of a 15-second load test _____ VDC

    Battery voltage 1 minute after load test _____ VDC

    State of battery charge _____ percent

    Task completed _____

3. Three-minute charging test:

    Charging rate _____ amps

    Voltage before charge _____ VDC

    Voltage after 3 minutes _____ VDC

    Task completed _____

STUDENT SELF-EVALUATION

Check	Level	Competency	Comments
	4	Mastered task	
	3	Competent but need further help	
	2	Needed a lot of help	
	0	Did not understand the task	

INSTRUCTOR EVALUATION

Check	Level	Competency	Comments
	4	Mastered task	
	3	Competent but needs further help	
	2	Requires more training	
	0	Unable to perform task	

ONLINE TASKS

Use a search engine to access the following Web pages and see what they have to say about batteries:

1. AC Delco
2. Delphi Freedom
3. Universal Power Group
4. Delco Remy

STUDY TIPS

Identify five key points in Chapter 7. Try to be as brief as possible.

Key point 1 _____

Key point 2 _____

Key point 3 _____

Key point 4 _____

Key point 5 _____

8 Charging Systems

Objectives

After reading this chapter, you should be able to:
- Identify charging circuit components.
- Navigate a charging circuit schematic.
- Voltage drop test charging circuit wiring and components.
- Describe the construction of an alternator.
- Explain full-wave rectification.
- Full-field an alternator.
- Measure AC leakage in the charging circuit.
- Verify the performance of an alternator.
- Use Intelli-check™ to assess charging circuit performance.
- Disassemble and reassemble a Delco Remy 33/40 SI alternator.

PRACTICE QUESTIONS

1. When is a diode likely to conduct current?
    a. forward bias
    b. reverse bias
    c. neutral state
    d. full-fielded

2. Technician A says that the brushes on alternators carry small current loads, which means they last a long time. Technician B says that the brushes on older generators had to conduct the specified current load of the alternator. Who is correct?
    a. Technician A only
    b. Technician B only
    c. both A and B
    d. neither A nor B

3. How many diodes are required on a three-phase alternator diode bridge to achieve full-wave rectification?
    a. one
    b. three
    c. six
    d. eighteen

4. In what type of truck charging/cranking circuit would you find a transformer/rectifier?
    a. 12-volt
    b. 12/24-volt
    c. 24-volt
    d. 42-volt

5. Which of the following test instruments would be preferred when verifying alternator current output?
   a. DC-voltmeter
   b. AC-voltmeter
   c. ammeter in series with alternator output cable
   d. ammeter with inductive pickup on alternator output cable

6. Technician A says that when full-fielding an alternator in chassis, maximum battery load should be drawn from the chassis battery banks. Technician B says that voltage spikes can be produced whenever an alternator is full-fielded. Who is correct?
   a. Technician A only
   b. Technician B only
   c. both A and B
   d. neither A nor B

7. When you are checking alternator AC leakage, what specific alternator component is being tested?
   a. brushes
   b. diode bridge
   c. field strength
   d. current output

8. When measuring alternator AC leakage, an AC voltage value of 0.8 V-AC is produced. What should you do?
   a. Adjust the voltage regulator.
   b. Replace the voltage regulator.
   c. Replace the alternator.
   d. Clean the alternator ground path.

9. When using Intelli-check on an alternator, the *low battery voltage* status light is illuminated. What should you do?
   a. Replace the alternator.
   b. Replace the diode bridge.
   c. Replace the starter motor.
   d. Test the charging circuit and recharge the batteries if necessary.

10. Technician A says that most current truck alternators use a delta configuration of diodes to achieve full-wave rectification. Technician B says that most truck alternators use solid-state voltage regulators. Who is correct?
    a. Technician A only
    b. Technician B only
    c. both A and B
    d. neither A nor B

JOB SHEET 8.1

Name _____ Station _____ Date _____

Test Alternator Output.

Performance Objective(s): Test a truck alternator output to specification using an AVR with an inductive current clamp and carbon pile tester.

ASE Education Foundation Correlation

This job sheet addresses the following ASE Education Foundation task(s):

V. Electrical/Electronic Systems

A. General
1. Research vehicle service information, including vehicle service history, service precautions, and technical service bulletins. (IMMR, TST, MTST) P-1.
2. Demonstrate knowledge of electrical/electronic series, parallel, and series-parallel circuits using principles of electricity (Ohm's Law). (IMMR, TST, MTST) P-1.
3. Demonstrate proper use of test equipment when measuring source voltage, voltage drop (including grounds), current flow, continuity, and resistance. (IMMR, TST, MTST) P-1.
4. Demonstrate knowledge of the causes and effects of shorts, grounds, opens, and resistance problems in electrical/electronic circuits. (IMMR) P-1.
4. Demonstrate knowledge of the causes and effects of shorts, grounds, opens, and resistance problems in electrical/electronic circuits; identify and locate faults in electrical/electronic circuits. (TST, MTST) P-1.
5. Use wiring diagrams to trace electrical/electronic circuits. (IMMR) P-1.
5. Use wiring diagrams during the diagnosis (troubleshooting) of electrical/electronic circuit problems. (TST, MTST) P-1.
9. Use appropriate electronic service tool(s) and procedures to check, record, and clear diagnostic codes; interpret digital multimeter (DMM) readings. (IMMR) P-2.
9. Use appropriate electronic service tool(s) and procedures to diagnose problems; check, record, and clear diagnostic codes; interpret digital multimeter (DMM) readings. (TST, MTST) P-2.
11. Identify electrical/electronic system components and configuration. (IMMR, TST, MTST) P-1.

B. Battery System
2. Confirm proper battery capacity for application; perform battery state-of-charge test; perform battery capacity test, determine needed action. (IMMR, TST, MTST) P-1.
4. Charge battery using appropriate method for battery type. (IMMR, TST, MTST) P-1.

D. Charging System
1. Identify and understand operation of the generator (alternator). (IMMR, TST, MTST) P-1.
2. Check instrument panel mounted voltmeters and/or indicator lamps. (IMMR) P-1.
2. Test instrument panel mounted voltmeters and/or indicator lamps; determine needed action. (TST, MTST) P-1.
3. Inspect generator (alternator) drive belt condition; check pulleys and tensioners for wear; check fans and mounting brackets; verify proper belt alignment. (IMMR) P-1.
3. Inspect, adjust, and/or replace generator (alternator) drive belt; check pulleys and tensioners for wear; check fans and mounting brackets; verify proper belt alignment; determine needed action. (TST, MTST) P-1.
4. Inspect cables, wires, and connectors in the charging circuit. (IMMR) P-1.
4. Inspect cables, wires, and connectors in the charging circuit; determine needed action. (TST, MTST) P-1.
5. Perform charging system voltage and amperage output tests; perform AC ripple test. (IMMR) P-1.
5. Perform charging system voltage and amperage output tests; perform AC ripple test; determine needed action. (TST, MTST) P-1.

Tools and Materials: A shop AVR equipped with an inductive current clamp pickup, a carbon pile load tester, and a truck with a functional electrical system.

Protective Clothing: Standard shop apparel including coveralls or shop coat, safety glasses, and safety footwear.

PROCEDURE
Ensure that the shop LOTO is observed.

The procedure described here is suitable for testing alternator output if a standard heavy-duty alternator is used. For systems that use TR (electronic 12/24 V) alternators, use the OEM test instructions.

1. Verify that the vehicle batteries are fully charged.

   Task completed _____

2. Connect a carbon pile tester across the battery terminals. Also connect a voltmeter across the battery terminals and verify the battery voltage.

   Battery voltage _____

   Task completed _____

3. Connect an inductive pickup ammeter clamp over the cable that connects the alternator to the starter solenoid BAT terminal. Ensure that the arrow on the ammeter clamp points away from the alternator (this will be the direction of current flow when the alternator is working), that is, toward the solenoid BAT terminal.

   Task completed _____

4. Ensure that the truck engine can be run safely (wheels blocked, exhaust extractors in place, and the transmission in neutral).

   Task completed _____

5. Start the truck engine and run under no load at its rated rpm. Now use the carbon pile to load down the electrical system. Load down the carbon pile while observing the voltmeter. Continue to do so until battery voltage reads 12.7 V. Record the current value. Unload the carbon pile and idle down the engine.

   Peak current: _____ amps

   Task completed _____

6. Check the alternator current rating specification. The peak current recorded in the previous step should be within 10 percent of the alternator rating. So, a 130-amp-rated alternator should have produced at least 115 amps in the step 5 test. If it does not, the alternator should be repaired or replaced.

   Rated current: _____ amps

   Alternator OK _____

   Task completed _____

STUDENT SELF-EVALUATION

Check	Level	Competency	Comments
	4	Mastered task	
	3	Competent but need further help	
	2	Needed a lot of help	
	0	Did not understand the task	

INSTRUCTOR EVALUATION

Check	Level	Competency	Comments
	4	Mastered task	
	3	Competent but needs further help	
	2	Requires more training	
	0	Unable to perform task	

JOB SHEET 8.2

Name _____ Station _____ Date _____

Test the Charging Circuit.

Performance Objective(s): Test a truck charging circuit to specification.

ASE Education Foundation Correlation

This job sheet addresses the following ASE Education Foundation task(s):

V. Electrical/Electronic Systems
A. General
1. Research vehicle service information, including vehicle service history, service precautions, and technical service bulletins. (IMMR, TST, MTST) P-1.
2. Demonstrate knowledge of electrical/electronic series, parallel, and series-parallel circuits using principles of electricity (Ohm's Law). (IMMR, TST, MTST) P-1.
3. Demonstrate proper use of test equipment when measuring source voltage, voltage drop (including grounds), current flow, continuity, and resistance. (IMMR, TST, MTST) P-1.
4. Demonstrate knowledge of the causes and effects of shorts, grounds, opens, and resistance problems in electrical/electronic circuits. (IMMR) P-1.
4. Demonstrate knowledge of the causes and effects of shorts, grounds, opens, and resistance problems in electrical/electronic circuits; identify and locate faults in electrical/electronic circuits. (TST, MTST) P-1.
5. Use wiring diagrams to trace electrical/electronic circuits. (IMMR) P-1.
5. Use wiring diagrams during the diagnosis (troubleshooting) of electrical/electronic circuit problems. (TST, MTST) P-1.
9. Use appropriate electronic service tool(s) and procedures to check, record, and clear diagnostic codes; interpret digital multimeter (DMM) readings. (IMMR) P-2.
9. Use appropriate electronic service tool(s) and procedures to diagnose problems; check, record, and clear diagnostic codes; interpret digital multimeter (DMM) readings. (TST, MTST) P-2.
11. Identify electrical/electronic system components and configuration. (IMMR, TST, MTST) P-1.

B. Battery System
2. Confirm proper battery capacity for application; perform battery state-of-charge test; perform battery capacity test, determine needed action. (IMMR, TST, MTST) P-1.
3. Inspect battery, battery cables, connectors, battery boxes, mounts, and hold-downs; determine needed action. (IMMR, TST, MTST) P-1.
4. Charge battery using appropriate method for battery type. (IMMR, TST, MTST) P-1.

D. Charging System
1. Identify and understand operation of the generator (alternator). (IMMR, TST, MTST) P-1.
2. Check instrument panel mounted voltmeters and/or indicator lamps. (IMMR) P-1.
2. Test instrument panel mounted voltmeters and/or indicator lamps; determine needed action. (TST, MTST) P-1.
3. Inspect generator (alternator) drive belt condition; check pulleys and tensioners for wear; check fans and mounting brackets; verify proper belt alignment. (IMMR) P-1.
3. Inspect, adjust, and/or replace generator (alternator) drive belt; check pulleys and tensioners for wear; check fans and mounting brackets; verify proper belt alignment; determine needed action. (TST, MTST) P-1.
4. Inspect cables, wires, and connectors in the charging circuit. (IMMR) P-1.
4. Inspect cables, wires, and connectors in the charging circuit; determine needed action. (TST, MTST) P-1.
5. Perform charging system voltage and amperage output tests; perform AC ripple test. (IMMR) P-1.
5. Perform charging system voltage and amperage output tests; perform AC ripple test; determine needed action. (TST, MTST) P-1.

E. Lighting Systems

1. Inspect for brighter-than-normal, intermittent, dim, or no-light operation; determine needed action. (IMMR) P-1.
1. Identify causes of brighter-than-normal, intermittent, dim, or no-light operation; determine needed action. (TST) P-1.
1. Diagnose causes of brighter-than-normal, intermittent, dim, or no-light operation; determine needed action. (MTST) P-1.

Tools and Materials: A shop AVR equipped with an inductive current clamp pickup, a carbon pile load tester, a DMM, and a truck with a functional electrical system.

Protective Clothing: Standard shop apparel including coveralls or shop coat, safety glasses, and safety footwear.

PROCEDURE
Ensure that the shop LOTO is observed.

The procedure described here is a generic one. Some OEMs require specific diagnostic test routines that may require connecting to the chassis data bus, an online SIS handshake, and OEM diagnostic software. Always consult the OEM test instructions before undertaking this procedure.

1. Check batteries. Visually inspect battery posts for looseness and corrosion and battery case for bulging and leaks. If possible, check battery electrolyte levels and specific gravity. Caution: Do not break open seals on maintenance-free, gelled electrolyte, and AGM batteries.

    Task completed _____

2. Test battery open-circuit voltage (OCV). Remove at least one battery cable from each battery to isolate each from the bank. Use a DMM to record V-Bat _____ V-DC. If the voltage reading is 12.5 V-DC or less and the truck has not been operated for 48 hours, charge the battery or batteries. If the OCV is over 12.8 V-DC, apply a moderate load (use a carbon pile set at 30 amps) for 1 minute to remove the surface charge. Then retest OCV _____ V-DC.

    Task completed _____

3. Load test the battery. Locate the CCA specification of the batteries used in the bank _____ CCA. Divide the CCA by 2 _____ ½ CCA. With each battery isolated from its bank, connect a carbon pile across the battery posts. Apply a load equivalent to ½ CCA for 15 seconds. After completion of each test, the battery voltage should be at 9.6 V-DC or higher. Record the results for each battery in the bank.

    Battery One _____ Good / Failed

    Battery Two _____ Good / Failed

    Battery Three _____ Good / Failed

    Battery Four _____ Good / Failed

    Providing the battery was known to be fully charged prior to the test, it can be regarded as requiring replacement if it fails this test.

    Task completed _____

4. Check alternator. Visually inspect the drive belts, pulleys, and mounting hardware. Start the engine and run to high idle rpm. Ensure that the batteries are fully charged and then measure the alternator output voltage and record _____ V-DC. The voltage reading produced should be 13.5 V-DC or greater but check OEM spec. If the voltage is lower than 13.5 V-DC when the engine is run at high idle, check for high-current draw.

    Task completed _____

5. Test alternator amperage. Ensure that the batteries are fully charged for this test. Connect an inductive pickup ammeter clamp over the cable that connects the alternator to the starter solenoid BAT terminal. Ensure that the arrow on the ammeter clamp points away from the alternator (this will be the direction of current flow when the alternator is working), that is, toward the solenoid BAT terminal. Keep the amp clamp around 12 inches away from any connection.

   Task completed _____

6. Start the truck engine and run under no load at its rated rpm. Now use the carbon pile to load down the electrical system. Load down the carbon pile while observing the voltmeter. Continue to do so until the battery voltage reads 12.7 V. Record the current value. Unload the carbon pile and idle down the engine.

   Peak current: _____ amps

   Task completed _____

7. Check the alternator current rating specification. The peak current recorded in the previous step should be within 10 percent of the alternator rating. So, a 130-amp-rated alternator should have produced at least 115 amps in the step 5 test. If it does not, the alternator should be repaired or replaced.

   Rated current: _____ amps

   Alternator OK _____

   Task completed _____

8. AC leakage test. Run the engine at 1,500 rpm or higher. Set a DMM at V-AC. Place one DMM test lead on the alternator insulated terminal and attach the other to chassis ground. Generally the maximum permitted AC leakage is 0.3 V-AC but check OEM specifications. Note that a high but within specification V-AC reading can identify an alternator that is about to fail.

   AC leakage: _____ V-AC

   Diode rectifier bridge OK _____

   Task completed _____

STUDENT SELF-EVALUATION

Check	Level	Competency	Comments
	4	Mastered task	
	3	Competent but need further help	
	2	Needed a lot of help	
	0	Did not understand the task	

INSTRUCTOR EVALUATION

Check	Level	Competency	Comments
	4	Mastered task	
	3	Competent but needs further help	
	2	Requires more training	
	0	Unable to perform task	

ONLINE TASKS

Use a search engine to access the following Web pages and find out what they have to say about truck alternators:

1. Delco Remy
2. Prestolite

STUDY TIPS

Identify five key points in Chapter 8. Try to be as brief as possible.

Key point 1 _____

Key point 2 _____

Key point 3 _____

Key point 4 _____

Key point 5 _____

9 Cranking Systems

Objectives

After reading this chapter, you should be able to:
- Identify the components in a truck cranking circuit.
- Explain the operating principles of magnetic switches, solenoids, and starter motors.
- Identify the control circuit components in a cranking circuit and define the role of the ECM in providing overcrank protection.
- Describe the operating principles of lightweight, planetary gear reduction starter motors.
- Test and troubleshoot a cranking circuit using voltage drop testing.
- Disassemble a heavy-duty truck starter motor.
- Test an armature for shorts using a growler.
- Test an armature for grounds and opens.
- Use a testlight to check out field coils.
- Outline the procedure required to rebuild a Remy 42MT starter motor.

PRACTICE QUESTIONS

1. What would be a typical current draw of a heavy-duty truck starter motor when cranking?
   a. 1.5 to 3 amps
   b. 100 to 200 amps
   c. 300 to 400 amps
   d. more than 1,200 amps

2. Before performing a test of a truck cranking circuit, which of the following must be performed first?
   a. Test the ignition circuit.
   b. Ensure that the batteries are fully charged.
   c. Check the charging circuit.
   d. Test the starter ground circuit.

3. When is an energized cranking motor capable of producing most torque?
   a. at close to stall rpm
   b. at maximum rpm
   c. after 15 seconds of cranking
   d. when it has completely warmed up

4. When testing a cranking circuit for voltage drops, you measure a 0.5-V-DC loss across the solenoid. Which of the following should you perform first?
   a. Clean the solenoid terminals.
   b. Replace the solenoid.
   c. Replace the starter motor.
   d. Replace the magnetic switch.

5. What is the maximum acceptable voltage drop across the insulated side of a truck cranking circuit?
   a. 0.1 V-DC
   b. 0.5 V-DC
   c. 1 V-DC
   d. 12.6 V-DC

6. Technician A says that an insulated cranking circuit cable that measures a 0.5-V-DC voltage drop should be replaced. Technician B says that a 0.2-V-DC drop across starter circuit cable connections is normal. Who is correct?
   a. Technician A only
   b. Technician B only
   c. both A and B
   d. neither A nor B

7. Which of the following is true of a series-wound starter motor?
   a. There is only one path for current flow.
   b. Maximum torque is produced at maximum rpm.
   c. Field coils rotate.
   d. No brushes are required.

8. Which of the following components rotates in a starter motor?
   a. field windings
   b. pull-in windings
   c. hold-in windings
   d. armature

9. Which of the following correctly explains the difference between starter solenoid pull-in and hold-in windings?
   a. Pull-in windings have more winding turns than hold-in windings.
   b. Pull-in windings are made of a heavier gauge than hold-in windings.
   c. Hold-in windings permit higher current flow.
   d. Hold-in windings are energized for less time during cranking.

10. What type of reduction gearing is used in a high-torque, lightweight starter motor such as a Delco-Remy 39 MT?
    a. crown and pinion gearset
    b. aluminum gear teeth
    c. planetary gearset
    d. viscous drive

JOB SHEET 9.1

Name _____ Station _____ Date _____

Major Electrical Component Wiring Circuit Analysis.

Performance Objective(s): Learn how the major electrical components in a truck wiring system are connected to each other. This exercise makes use of the knowledge you picked up in Chapters 5, 7, and 8, as well as this chapter.

ASE Education Foundation Correlation

This job sheet addresses the following ASE Education Foundation task(s):

V. Electrical/Electronic Systems
A. General
1. Research vehicle service information, including vehicle service history, service precautions, and technical service bulletins. (IMMR, TST, MTST) P-1.
2. Demonstrate knowledge of electrical/electronic series, parallel, and series-parallel circuits using principles of electricity (Ohm's Law). (IMMR, TST, MTST) P-1.
3. Demonstrate proper use of test equipment when measuring source voltage, voltage drop (including grounds), current flow, continuity, and resistance. (IMMR, TST, MTST) P-1.
4. Demonstrate knowledge of the causes and effects of shorts, grounds, opens, and resistance problems in electrical/electronic circuits. (IMMR) P-1.
4. Demonstrate knowledge of the causes and effects of shorts, grounds, opens, and resistance problems in electrical/electronic circuits; identify and locate faults in electrical/electronic circuits. (TST, MTST) P-1.
5. Use wiring diagrams to trace electrical/electronic circuits. (IMMR) P-1.
5. Use wiring diagrams during the diagnosis (troubleshooting) of electrical/electronic circuit problems. (TST, MTST) P-1.
7. Demonstrate knowledge of the function, operation, and testing of fusible links, circuit breakers, relays, solenoids, diodes, and fuses. (IMMR) P-1.
7. Demonstrate knowledge of the function, operation, and testing of fusible links, circuit breakers, relays, solenoids, diodes, and fuses; perform inspection and testing; determine needed action. (TST, MTST) P-1.
8. Inspect, repair (including solder repair), and/or replace connectors, seals, terminal ends, and wiring; verify proper routing and securement. (IMMR) P-1.
8. Inspect, test, repair (including solder repair), and/or replace components, connectors, seals, terminal ends, harnesses, and wiring; verify proper routing and securement; determine needed action. (TST, MTST) P-1.
9. Use appropriate electronic service tool(s) and procedures to check, record, and clear diagnostic codes; interpret digital multimeter (DMM) readings. (IMMR) P-2.
9. Use appropriate electronic service tool(s) and procedures to diagnose problems; check, record, and clear diagnostic codes; interpret digital multimeter (DMM) readings. (TST, MTST) P-2.

B. Battery System
1. Identify battery type and system configuration. (IMMR, TST, MTST) P-1.
2. Confirm proper battery capacity for application; perform battery state-of-charge test; perform battery capacity test, determine needed action. (IMMR, TST, MTST) P-1.
3. Inspect battery, battery cables, connectors, battery boxes, mounts, and hold-downs; determine needed action. (IMMR, TST, MTST) P-1.

C. Starting System
1. Demonstrate understanding of starter system operation. (IMMR, TST, MTST) P-1.
2. Perform starter circuit cranking voltage and voltage drop tests. (IMMR) P-1.
2. Perform starter circuit cranking voltage and voltage drop tests; determine needed action. (TST, MTST) P-1.

3. Inspect starter control circuit switches, relays, connectors, terminals, wires, and harnesses (including over-crank protection). (IMMR) P-1.
3. Inspect and test starter control circuit switches (key switch, push button, and/or magnetic switch), relays, connectors, terminals, wires, and harnesses (including over-crank protection); determine needed action. (TST, MST) P-1.

Tools and Materials: A heavy-duty truck with a functional electrical system.

Protective Clothing: Standard shop apparel including coveralls or shop coat, safety glasses, and safety footwear.

PROCEDURE
Ensure that the shop LOTO is observed.

For this exercise, do not use the OEM wiring schematic. Reference Figure 9–1 and then use a heavy-duty truck to connect the main components of the charging and cranking circuits.

Task completed _____

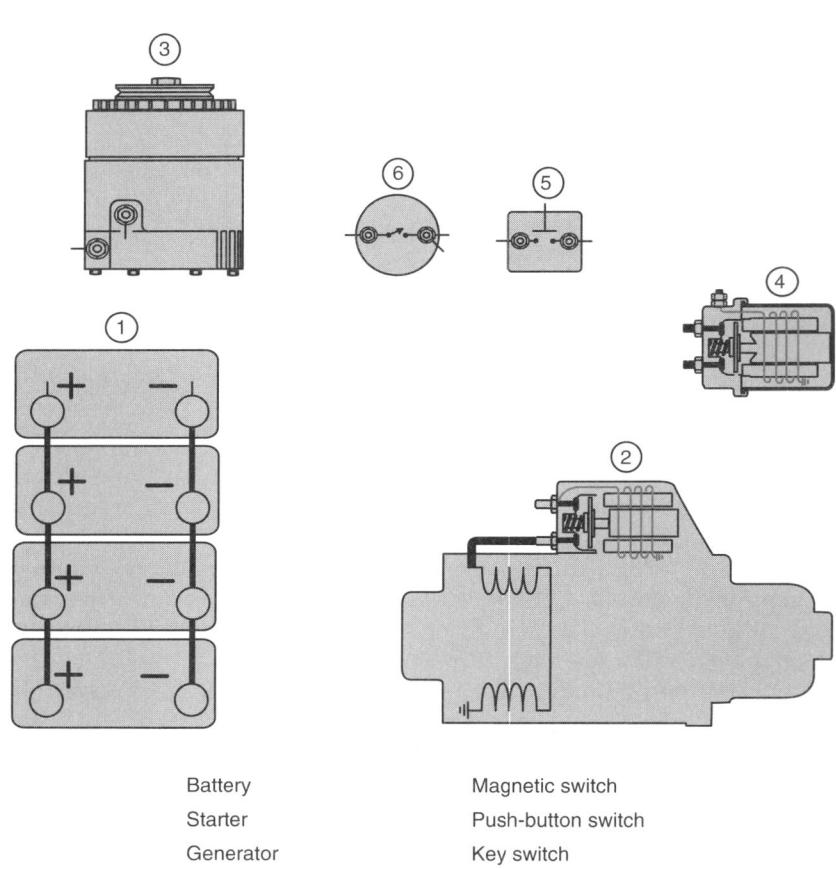

Battery
Starter
Generator
Magnetic switch
Push-button switch
Key switch

FIGURE 9–1 Connect these electrical components. Identify each component by number and use a different color for positive, ground, and control circuits.

STUDENT SELF-EVALUATION

Check	Level	Competency	Comments
	4	Mastered task	
	3	Competent but need further help	
	2	Needed a lot of help	
	0	Did not understand the task	

INSTRUCTOR EVALUATION

Check	Level	Competency	Comments
	4	Mastered task	
	3	Competent but needs further help	
	2	Requires more training	
	0	Unable to perform task	

JOB SHEET 9.2

Name _____ Station _____ Date _____

Perform Voltage Drop Tests on Cranking Circuit.

Performance Objective(s): Perform voltage drop tests on an energized truck cranking circuit.

ASE Education Foundation Correlation

This job sheet addresses the following ASE Education Foundation task(s):

V. Electrical/Electronic Systems
A. General
1. Research vehicle service information, including vehicle service history, service precautions, and technical service bulletins. (IMMR, TST, MTST) P-1.
2. Demonstrate knowledge of electrical/electronic series, parallel, and series- parallel circuits using principles of electricity (Ohm's Law). (IMMR, TST, MTST) P-1.
3. Demonstrate proper use of test equipment when measuring source voltage, voltage drop (including grounds), current flow, continuity, and resistance. (IMMR, TST, MTST) P-1.
4. Demonstrate knowledge of the causes and effects of shorts, grounds, opens, and resistance problems in electrical/electronic circuits. (IMMR) P-1.
4. Demonstrate knowledge of the causes and effects of shorts, grounds, opens, and resistance problems in electrical/electronic circuits; identify and locate faults in electrical/electronic circuits. (TST, MTST) P-1.
5. Use wiring diagrams to trace electrical/electronic circuits. (IMMR) P-1.
5. Use wiring diagrams during the diagnosis (troubleshooting) of electrical/electronic circuit problems. (TST, MTST) P-1.
9. Use appropriate electronic service tool(s) and procedures to check, record, and clear diagnostic codes; interpret digital multimeter (DMM) readings. (IMMR) P-2.
9. Use appropriate electronic service tool(s) and procedures to diagnose problems; check, record, and clear diagnostic codes; interpret digital multimeter (DMM) readings. (TST, MTST) P-2.
10. Check for malfunctions caused by faults in the data bus communications network. (IMMR, TST) P-2.
10. Diagnose faults in the data bus communications network; determine needed action. (MSTS) P-2.

C. Starting System
1. Demonstrate understanding of starter system operation. (IMMR, TST, MTST) P-1.
2. Perform starter circuit cranking voltage and voltage drop tests. (IMMR) P-1.
2. Perform starter circuit cranking voltage and voltage drop tests; determine needed action. (TST, MTST) P-1.

Tools and Materials: A DMM and a truck with a functional electrical circuit.

Protective Clothing: Standard shop apparel including coveralls or shop coat, safety glasses, and safety footwear.

PROCEDURE
Ensure that the shop LOTO is observed.

This is a critical diagnostic test that the technician should perform many times over to be completely familiar with it. Though the testing of truck cranking circuits is outlined in this job sheet, voltage drop tests may be performed anywhere on electrical circuits and components. Unlike a resistance test, voltage drop testing verifies the performance of circuits, wiring, terminals, and components under load so it can be regarded as a much more accurate assessment.

Set the DMM to the V-DC mode. It is best to use an auto-ranging DMM for this test, but if it is not available, a DMM capable of accurately reading voltage values below 1 V while in the 18 V scale must be used. Ensure that the engine will not start during the test. This requires no-fueling hydromechanical engines by actuating the no-fuel lever (Mack, International, DDC) or disconnecting the fuel solenoid (Caterpillar and Cummins). With most electronically managed diesel engines, the easiest way to no-fuel the engine for this test is usually to pull the EUI fuses (early generation) or program the ECM for no-start, diagnostic test mode (current). Whatever the system, always check the OEM literature for the required procedure.

1. With the volts set at V-DC, clamp the positive DMM probe to the battery positive post and the negative DMM probe to the starter solenoid BAT terminal. Crank the engine and record the voltage drop value.

    Voltage drop _____

    Task completed _____

2. Check the voltage drop you measured with the maximum specification. If it exceeds the maximum specification, try to identify the location of the high resistance by performing voltage drop tests from the battery post to the battery terminal, from the solenoid terminal to the solenoid post, and through the cable itself. Make sure you test each cable-end connector.

    Task completed _____

3. Now check for voltage losses through the ground side of the cranking circuit. Use the same DMM settings as in the previous test. Clamp the DMM positive probe on the starter ground post and the negative probe on the battery negative post. Crank the engine and record the voltage value.

    Voltage drop _____

    Task completed _____

4. If the voltage drop exceeds the maximum specification, try to identify the location of the high resistance by isolating connections in the ground circuit as in test 3. Remember that excessive heat is a good tattletale for high resistance in a circuit.

    Task completed _____

STUDENT SELF-EVALUATION

Check	Level	Competency	Comments
	4	Mastered task	
	3	Competent but need further help	
	2	Needed a lot of help	
	0	Did not understand the task	

INSTRUCTOR EVALUATION

Check	Level	Competency	Comments
	4	Mastered task	
	3	Competent but needs further help	
	2	Requires more training	
	0	Unable to perform task	

JOB SHEET 9.3

Name _____ Station _____ Date _____

Test the Cranking and Charging Circuits.

Performance Objective(s): Test truck cranking and charging circuits to specification. This job sheet replicates shop procedures used in job sheets located in Chapters 7 (batteries) and 8 (charging circuits) but adds cranking circuits.

ASE Education Foundation Correlation

This job sheet addresses the following ASE Education Foundation task(s):

V. Electrical/Electronic Systems

A. General
1. Research vehicle service information, including vehicle service history, service precautions, and technical service bulletins. (IMMR, TST, MTST) P-1.
2. Demonstrate knowledge of electrical/electronic series, parallel, and series-parallel circuits using principles of electricity (Ohm's Law). (IMMR, TST, MTST) P-1.
3. Demonstrate proper use of test equipment when measuring source voltage, voltage drop (including grounds), current flow, continuity, and resistance. (IMMR, TST, MTST) P-1.
8. Inspect, repair (including solder repair), and/or replace connectors, seals, terminal ends, and wiring; verify proper routing and securement. (IMMR) P-1.
8. Inspect, test, repair (including solder repair), and/or replace components, connectors, seals, terminal ends, harnesses, and wiring; verify proper routing and securement; determine needed action. (TST, MTST) P-1.
9. Use appropriate electronic service tool(s) and procedures to check, record, and clear diagnostic codes; interpret digital multimeter (DMM) readings. (IMMR) P-2.
9. Use appropriate electronic service tool(s) and procedures to diagnose problems; check, record, and clear diagnostic codes; interpret digital multimeter (DMM) readings. (TST, MTST) P-2.

B. Battery System
1. Identify battery type and system configuration. (IMMR, TST, MTST) P-1.
2. Confirm proper battery capacity for application; perform battery state-of-charge test; perform battery capacity test, determine needed action. (IMMR, TST, MTST) P-1.
3. Inspect battery, battery cables, connectors, battery boxes, mounts, and hold-downs; determine needed action. (IMMR, TST, MTST) P-1.
4. Charge battery using appropriate method for battery type. (IMMR, TST, MTST) P-1.
6. Identify low voltage disconnect (LVD) systems. (IMMR) P-2.
6. Check low voltage disconnect (LVD) systems; determine needed action. (TST, MTST) P-2.

C. Starting System
1. Demonstrate understanding of starter system operation. (IMMR, TST, MTST) P-1.
2. Perform starter circuit cranking voltage and voltage drop tests. (IMMR) P-1.
2. Perform starter circuit cranking voltage and voltage drop tests; determine needed action. (TST, MTST) P-1.
3. Inspect starter control circuit switches, relays, connectors, terminals, wires, and harnesses including (over-crank protection). (IMMR) P-1.
3. Inspect and test starter control circuit switches (key switch, push button, and/or magnetic switch), relays, connectors, terminals, wires, and harnesses (including over-crank protection); determine needed action. (TST, MST) P-1.

D. Charging System
1. Identify and understand operation of the generator (alternator). (IMMR, TST, MTST) P-1.
2. Check instrument panel mounted voltmeters and/or indicator lamps. (IMMR) P-1.
2. Test instrument panel mounted voltmeters and/or indicator lamps; determine needed action. (TST, MTST) P-1.
3. Inspect generator (alternator) drive belt condition; check pulleys and tensioners for wear; check fans and mounting brackets; verify proper belt alignment. (IMMR) P-1.

3. Inspect, adjust, and/or replace generator (alternator) drive belt; check pulleys and tensioners for wear; check fans and mounting brackets; verify proper belt alignment; determine needed action. (TST, MTST) P-1.
4. Inspect cables, wires, and connectors in the charging circuit. (IMMR) P-1.
4. Inspect cables, wires, and connectors in the charging circuit; determine needed action. (TST, MTST) P-1.
5. Perform charging system voltage and amperage output tests; perform AC ripple test. (IMMR) P-1.
5. Perform charging system voltage and amperage output tests; perform AC ripple test; determine needed action. (TST, MTST) P-1.

Tools and Materials: A shop AVR equipped with an inductive current clamp pickup, a carbon pile load tester, a DMM, and a truck with a functional electrical system.

Protective Clothing: Standard shop apparel including coveralls or shop coat, safety glasses, and safety footwear.

PROCEDURE
Ensure that the shop LOTO is observed.

The procedure described here is a generic one. Some OEMs require specific diagnostic test routines that may require connecting to the chassis data bus, an online SIS handshake, and OEM diagnostic software. Always consult the OEM test instructions before undertaking this procedure.

1. Check the batteries. Visually inspect the battery posts for looseness and corrosion and the battery case for bulging and leaks. If possible, check battery electrolyte levels and specific gravity. Caution: Do not break open seals on maintenance-free, gelled electrolyte, and AGM batteries.

    Task completed _____

2. Test battery open-circuit voltage (OCV). Remove at least one battery cable from each battery to isolate each from the bank. Use a DMM to record V-Bat _____ V-DC. If the voltage reading is 12.5 V-DC or less and the truck has not been operated for 48 hours, charge the battery or batteries. If the OCV is over 12.8 V-DC, apply a moderate load (use a carbon pile set at 30 amps) for 1 minute to remove the surface charge. Then retest OCV _____ V-DC.

    Task completed _____

3. Load test battery. Locate the CCA specification of the batteries used in the bank _____ CCA. Divide the CCA by 2 _____ ½ CCA. With each battery isolated from its bank, connect a carbon pile across the battery posts. Apply a load equivalent to ½ CCA for 15 seconds. After completion of each test, the battery voltage should be at 9.6 V-DC or higher. Record the results for each battery in the bank.

    Battery One _____ Good / Failed
    Battery Two _____ Good / Failed
    Battery Three _____ Good / Failed
    Battery Four _____ Good / Failed

    Providing the battery was known to be fully charged prior to the test, it can be regarded as requiring replacement if it fails this test.

    Task completed _____

4. Check the alternator. Visually inspect the drive belts, pulleys, and mounting hardware. Start the engine and run to high idle rpm. Ensure that the batteries are fully charged and then measure the alternator output voltage and record _____ V-DC. The voltage reading produced should be 13.5 V-DC or greater but check OEM spec. If the voltage is lower than 13.5 V-DC when the engine is run at high idle, check for high current draw.

    Task completed _____

5. Test alternator amperage. Ensure that the batteries are fully charged for this test. Connect an inductive pickup ammeter clamp over the cable that connects the alternator to the starter solenoid BAT terminal. Ensure that the arrow on the ammeter clamp points away from the alternator (this will be the direction of current flow when the alternator is working), that is, toward the solenoid BAT terminal. Keep the amp clamp around 12 inches (300 mm) away from any connection.

   Task completed _____

6. Start the truck engine and run under no load at its rated rpm. Now use the carbon pile to load down the electrical system. Load down the carbon pile while observing the voltmeter. Continue to do so until battery voltage reads 12.7 V. Record the current value. Unload the carbon pile and idle down the engine.

   Peak current: _____ amps

   Task completed _____

7. Check the alternator current rating specification. The peak current recorded in the previous step should be within 10 percent of the alternator rating. So, a 130-amp-rated alternator should have produced at least 115 amps in the step 5 test. If it does not, the alternator should be repaired or replaced.

   Rated current: _____ amps   Alternator OK _____

   Task completed _____

8. AC leakage test. Run the engine at 1,500 rpm or higher. Set a DMM at V-AC. Place on DMM test lead on the alternator insulated terminal and the other to chassis ground. Generally the maximum permitted AC leakage is 0.3 V-AC but check OEM specifications. Note that a high but within specification V-AC reading can identify an alternator that is about to fail.

   AC leakage: _____ V-AC

   Diode rectifier bridge OK _____   Task completed _____

9. Test starter motor draw. Place an AVR amp clamp over the insulated starter cable, ensuring that the direction arrow is correctly set. Consult OEM service literature so that you use the correct procedure. You may be required to disable a sensor(s) to ensure a no-start. Use a DMM set to measure 2 V-DC or less. Crank the engine. You should observe an initial current surge (may exceed 500 amps) followed by a stabilized reading. Crank for 10 seconds and record the stabilized reading _____ amps current draw.

   Task completed _____

10. Cranking circuit voltage drop tests. Use a DMM set to accurately measure 2 V-DC or less. Place one DMM test lead on the starter motor positive terminal and the other on the battery positive stud. Crank the engine and record the reading. This should typically be less than 0.5 V-DC but check OEM specification _____ V-DC cranking motor insulated circuit voltage drop. Place one DMM test lead on the starter ground stud and the other on the battery negative stud. Crank the engine and record the reading. Again this should typically be less than 0.5 V-DC but check OEM specification. Readings close to the maximum spec can indicate an imminent problem. _____ V-DC ground circuit voltage drop.

    Task completed _____

11. Cab insulated circuit voltage drop test. Locate the cab power supply positive terminal. This is usually located to the fuse/breaker box in the dash or on the power management module. Check OEM specifications and diagnostic routines. Use a DMM set to accurately measure 2 V-DC or less. Place one DMM test lead on the cab power supply positive terminal and the other on the battery positive stud. Turn the cab blower motor on full and all of the vehicle lights and accessories. The reading should typically be less than 0.2 V-DC but check OEM specifications. _____ V-DC cab insulated circuit voltage drop.

    Task completed _____

12. Cab ground circuit voltage drop test. Use a DMM set to accurately measure 2 V-DC or less. Place one DMM test lead on a known good ground on the dash and the other on the battery negative stud. Turn the cab blower motor on full and all of the vehicle lights and accessories. The reading should typically be less than 0.3 V-DC but check OEM specifications. _____ V-DC cab ground circuit voltage drop.

Task completed _____

STUDENT SELF-EVALUATION

Check	Level	Competency	Comments
	4	Mastered task	
	3	Competent but need further help	
	2	Needed a lot of help	
	0	Did not understand the task	

INSTRUCTOR EVALUATION

Check	Level	Competency	Comments
	4	Mastered task	
	3	Competent but needs further help	
	2	Requires more training	
	0	Unable to perform task	

ONLINE TASKS

Use a search engine to access the following Web pages and note what each has to say about the starter motors they build:

1. Delco Remy
2. Prestolite
3. Denso

STUDY TIPS

Identify five key points in Chapter 9. Try to be as brief as possible.

Key point 1 _____

Key point 2 _____

Key point 3 _____

Key point 4 _____

Key point 5 _____

10 Chassis Electrical Circuits

Objectives

After reading this chapter, you should be able to:

- Describe how a light bulb functions.
- Explain the operating principles of halogen, LED, and high-intensity discharge (HID) lamps.
- Describe the function of the reflector and lens in a headlamp assembly.
- Aim truck headlights.
- Troubleshoot lighting circuit malfunctions.
- Describe the operation of typical truck auxiliary equipment.
- Explain how a trailer electrical plug and connector are connected.
- Outline the operating principles of truck instrument cluster components.
- Diagnose and repair some typical truck instrument cluster failures.
- Explain the function and operation of warning and shutdown systems.
- Identify the types of circuit protection used in truck electrical systems, including fuses and cycling and noncycling circuit breakers.
- Describe the procedure and material required to solder a pair of copper wires.
- Outline the procedure required to quickly check out a truck electrical system.

PRACTICE QUESTIONS

1. What lights would be located on either side of a truck cab roof and at upper extremities of a van trailer?
   a. identification lights
   b. clearance lights
   c. fog lights
   d. intermediate turn signal lights

2. What component in an incandescent light bulb emits light?
   a. contact terminals
   b. inert gas
   c. tungsten filament
   d. argon

3. What federal legislation covers the lighting standards required by all U.S.-built highway trucks?
   a. FMVSS 105
   b. FMVSS 108
   c. FMVSS 121
   d. FMVSS 164

4. Which of the following types of light element would usually only be found in a headlight assembly?
   a. LED
   b. halogen
   c. fluorescent
   d. tungsten element

5. What are high-intensity discharge (HID) headlamps often known as?
    a. sodium lights
    b. fluorescent lights
    c. xenon lights
    d. infrared lights

6. How much voltage is required to maintain the arc in an HID light after the arc has been ignited?
    a. 20,000 volts
    b. 85 volts
    c. 42 volts
    d. 12 volts

7. Which of the following conditions is most likely to cause a failure of an HID light element?
    a. oxidation of the electrodes
    b. road shock and vibration
    c. system voltage spikes
    d. bad grounding

8. Why are LED lighting systems becoming so popular in truck and trailer lighting systems?
    a. greater longevity
    b. less current use
    c. quicker illumination
    d. all of the above

9. When adjusting the aim of a truck headlamp assembly using a wall adjustment screen, how far away from the wall should the truck be positioned?
    a. 8 feet
    b. 16 feet
    c. 25 feet
    d. 100 feet

10. In a standard seven-wire trailer connector cord, what does the brown color code represent?
    a. auxiliary
    b. brakes
    c. ground
    d. tail and license plates

JOB SHEET 10.1

Name _____ Station _____ Date _____

Construct a Relay-Controlled Electrical Circuit.

Performance Objective(s): Build up a relay switched electrical circuit.

ASE Education Foundation Correlation
This job sheet addresses the following ASE Education Foundation task(s):

V. Electrical/Electronic Systems
  A. General
  1. Research vehicle service information, including vehicle service history, service precautions, and technical service bulletins. (IMMR, TST, MTST) P-1.
  2. Demonstrate knowledge of electrical/electronic series, parallel, and series-parallel circuits using principles of electricity (Ohm's Law). (IMMR, TST, MTST) P-1.

  C. Starting System
  3. Inspect starter control circuit switches, relays, connectors, terminals, wires, and harnesses (including over-crank protection). (IMMR) P-1.
  3. Inspect and test starter control circuit switches (key switch, push button, and/or magnetic switch), relays, connectors, terminals, wires, and harnesses (including over-crank protection); determine needed action. (TST, MST) P-1.

Tools and Materials: Any standard SAE relay, a battery, switch, headlamp, and interconnecting wiring. Identify the function of each relay terminal.

Protective Clothing: Standard shop apparel including coveralls or shop coat, safety glasses, and safety footwear.

PROCEDURE
Ensure that the shop LOTO is observed.

Use Figure 10–1 as a model to build a relay-switched headlamp circuit and then fill in the data fields as indicated on the diagram.

Task completed _____

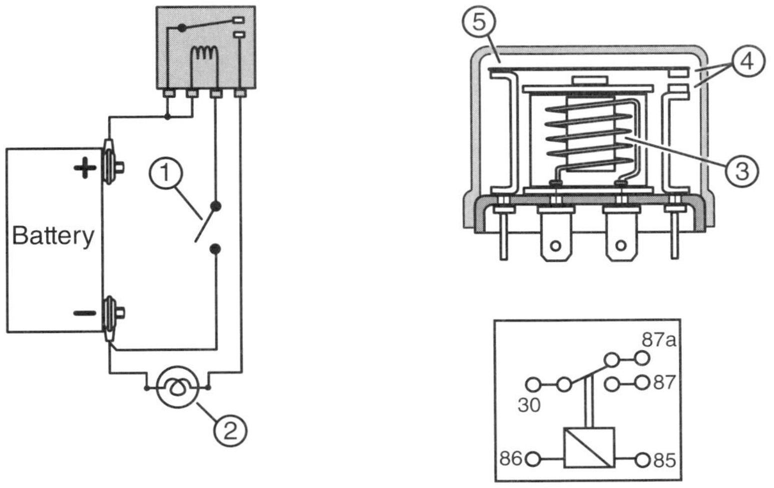

1. _____   2. _____   3. _____   4. _____   5. _____

FIGURE 10–1 Relays: Identify the components. Cross-reference the relay terminal codes to the newer SAE 1-5 codes.

STUDENT SELF-EVALUATION

Check	Level	Competency	Comments
	4	Mastered task	
	3	Competent but need further help	
	2	Needed a lot of help	
	0	Did not understand the task	

INSTRUCTOR EVALUATION

Check	Level	Competency	Comments
	4	Mastered task	
	3	Competent but needs further help	
	2	Requires more training	
	0	Unable to perform task	

JOB SHEET 10.2

Name _____ Station _____ Date _____

Perform Voltage Drop Test on Insulated Side of the Cranking Circuit.

Performance Objective(s): Perform a sequential voltage drop test on the insulated side of a cranking circuit of a commercial vehicle.

ASE Education Foundation Correlation

This job sheet addresses the following ASE Education Foundation task(s):

V. Electrical/Electronic Systems
  A. General
  1. Research vehicle service information, including vehicle service history, service precautions, and technical service bulletins. (IMMR, TST, MTST) P-1.
  2. Demonstrate knowledge of electrical/electronic series, parallel, and series-parallel circuits using principles of electricity (Ohm's Law). (IMMR, TST, MTST) P-1.

  C. Starting System
  2. Perform starter circuit cranking voltage and voltage drop tests. (IMMR) P-1.
  2. Perform starter circuit cranking voltage and voltage drop tests; determine needed action. (TST, MTST) P-1.

Tools and Materials: Truck with a functional cranking circuit and an accurate DMM.

Protective Clothing: Standard shop apparel including coveralls or shop coat, safety glasses, and safety footwear.

PROCEDURE
Ensure that the shop LOTO is observed.

Reference the procedure outlined in Chapter 10 under the headings Voltage Drop Test and Performing a Voltage Drop Test. Set a DMM to read voltage values at 2 V-DC or less (not required with an auto-ranging DMM), and fill in the following fields. Valid readings can only be obtained while the engine is being cranked. Note that the positive DMM test probe should be placed on the component listed first on the table that follows.

Insulated Circuit Components	Voltage Drop
Bat + post, to + Bat cable clamp/terminal	
+ Bat cable clamp/terminal to + cable clamp	
+ Cable clamp to + starter terminal post	

Task completed _____

STUDENT SELF-EVALUATION

Check	Level	Competency	Comments
	4	Mastered task	
	3	Competent but need further help	
	2	Needed a lot of help	
	0	Did not understand the task	

INSTRUCTOR EVALUATION

Check	Level	Competency	Comments
	4	Mastered task	
	3	Competent but needs further help	
	2	Requires more training	
	0	Unable to perform task	

JOB SHEET 10.3

Name _____ Station _____ Date _____

Test the Cranking, Charging, and Cab Electrical Circuits.

Performance Objective(s): Test truck cranking and charging circuits to specification. This job sheet replicates shop procedures used in job sheets located in Chapters 7 (batteries), 8 (charging circuits), and 9 (cranking circuits) by adding cab electrical tests.

ASE Education Foundation Correlation

This job sheet addresses the following ASE Education Foundation task(s):

V. Electrical/Electronic Systems

A. General
1. Research vehicle service information, including vehicle service history, service precautions, and technical service bulletins. (IMMR, TST, MTST) P-1.
2. Demonstrate knowledge of electrical/electronic series, parallel, and series-parallel circuits using principles of electricity (Ohm's Law). (IMMR, TST, MTST) P-1.
3. Demonstrate proper use of test equipment when measuring source voltage, voltage drop (including grounds), current flow, continuity, and resistance. (IMMR, TST, MTST) P-1.

B. Battery System
1. Identify battery type and system configuration. (IMMR, TST, MTST) P-1.
2. Confirm proper battery capacity for application; perform battery state-of-charge test; perform battery capacity test, determine needed action. (IMMR, TST, MTST) P-1.
3. Inspect battery, battery cables, connectors, battery boxes, mounts, and hold-downs; determine needed action. (IMMR, TST, MTST) P-1.
4. Charge battery using appropriate method for battery type. (IMMR, TST, MTST) P-1.
5. Jump-start vehicle using a booster battery and jumper cables or using an appropriate auxiliary power supply. (IMMR, TST, MTST) P-1.
6. Identify low voltage disconnect (LVD) systems. (IMMR) P-1.
6. Check low voltage disconnect (LVD) systems; determine needed action. (TST, MTST) P-1.

C. Starting System
1. Demonstrate understanding of starter system operation. (IMMR, TST, MTST) P-1.
2. Perform starter circuit cranking voltage drop tests; determine voltage and voltage drop tests. (IMMR) P-1.
2. Perform starter circuit cranking voltage and needed action. (TST, MTST) P-1.
3. Inspect starter control circuit switches, relays, connectors, terminals, wires, and harnesses (including over-crank protection). (IMMR) P-1.
3. Inspect and test starter control circuit switches (key switch, push button, and/or magnetic switch), relays, connectors, terminals, wires, and harnesses (including over-crank protection); determine needed action. (TST, MST) P-1.

D. Charging System
1. Identify and understand operation of the generator (alternator). (IMMR, TST, MTST) P-1.
2. Check instrument panel mounted voltmeters and/or indicator lamps. (IMMR) P-1.
2. Test instrument panel mounted voltmeters and/or indicator lamps; determine needed action. (TST, MTST) P-1.
3. Inspect generator (alternator) drive belt condition; check pulleys and tensioners for wear; check fans and mounting brackets; verify proper belt alignment. (IMMR) P-1.
3. Inspect, adjust, and/or replace generator (alternator) drive belt; check pulleys and tensioners for wear; check fans and mounting brackets; verify proper belt alignment; determine needed action. (TST, MTST) P-1.
4. Inspect cables, wires, and connectors in the charging circuit; determine needed action. (IMMR, TST, MTST) P-1.

5. Perform charging system voltage and amperage output tests; perform AC ripple test. (IMMR) P-1.
5. Perform charging system voltage and amperage output tests; perform AC ripple test; determine needed action. (TST, MTST) P-1.

Tools and Materials: A shop AVR equipped with an inductive current clamp pickup, a carbon pile load tester, a DMM, and a truck with a functional electrical system.

Protective Clothing: Standard shop apparel including coveralls or shop coat, safety glasses, and safety footwear.

PROCEDURE
Ensure that the shop LOTO is observed.

The procedure described here is a generic one. Some OEMs require specific diagnostic test routines that may require connecting to the chassis data bus, an online SIS handshake, and OEM diagnostic software. Always consult the OEM test instructions before undertaking this procedure.

1. Check the batteries. Visually inspect the battery posts for looseness and corrosion and the battery case for bulging and leaks. If possible, check battery electrolyte levels and specific gravity. Caution: Do not break open seals on maintenance-free, gelled electrolyte, and AGM batteries.

    Task completed _____

2. Test battery open circuit voltage (OCV). Remove at least one battery cable from each battery to isolate each from the bank. Use a DMM to record V-Bat _____ V-DC. If the voltage reading is 12.5 V-DC or less and the truck has not been operated for 48 hours, charge the battery or batteries. If the OCV is over 12.8 V-DC, apply a moderate load (use a carbon pile set at 30 amps) for 1 minute to remove the surface charge. Then retest OCV _____ V-DC.

    Task completed _____

3. Load test battery. Locate the CCA specification of the batteries used in the bank _____ CCA. Divide the CCA by 2 _____ ½ CCA. With each battery isolated from its bank, connect a carbon pile across the battery posts. Apply a load equivalent to ½ CCA for 15 seconds. After completion of each test, the battery voltage should be at 9.6 V-DC or higher. Record the results for each battery in the bank.

    Battery One _____ Good / Failed

    Battery Two _____ Good / Failed

    Battery Three _____ Good / Failed

    Battery Four _____ Good / Failed

    Providing the battery was known to be fully charged prior to the test, it can be regarded as requiring replacement if it fails this test.

    Task completed _____

4. Check the alternator. Visually inspect the drive belts, pulleys, and mounting hardware. Start the engine and run to high idle rpm. Ensure that the batteries are fully charged and then measure the alternator output voltage and record _____ V-DC. The voltage reading produced should be 13.5 V-DC or greater but check OEM spec. If the voltage is lower than 13.5 V-DC when the engine is run at high idle, check for high current draw.

    Task completed _____

5. Test alternator amperage. Ensure that the batteries are fully charged for this test. Connect an inductive pickup ammeter clamp over the cable that connects the alternator to the starter solenoid BAT terminal. Ensure that the arrow on the ammeter clamp points away from the alternator (this will be the direction of current flow when the alternator is working), that is, toward the solenoid BAT terminal. Keep the amp clamp around 12 inches away from any connection.

    Task completed _____

6. Start the truck engine and run under no load at its rated rpm. Now use the carbon pile to load down the electrical system. Load down the carbon pile while observing the voltmeter. Continue to do so until battery voltage reads 12.7 V. Record the current value. Unload the carbon pile and idle down the engine.

   Peak current: _____ amps

   Task completed _____

7. Check the alternator current rating specification. The peak current recorded in the previous step should be within 10 percent of the alternator rating. So, a 130-amp-rated alternator should have produced at least 115 amps in the step 5 test. If it does not, the alternator should be repaired or replaced.

   Rated current: _____ amps

   Alternator OK _____

   Task completed _____

8. AC leakage test. Run the engine at 1,500 rpm or higher. Set a DMM at V-AC. Place on DMM test lead on the alternator insulated terminal and the other to chassis ground. Generally the maximum permitted AC leakage is 0.3 V-AC but check OEM specifications. Note that a high but within specification V-AC reading can identify an alternator that is about to fail.

   AC leakage: _____ V-AC

   Diode rectifier bridge OK _____

   Task completed _____

9. Test starter motor draw. Place an AVR amp clamp over the insulated starter cable, ensuring that the direction arrow is correctly set. Consult OEM service literature so that you use the correct procedure. You may be required to disable a sensor(s) to ensure a no-start. Use a DMM set to measure 2 V-DC or less. Crank the engine. You should observe an initial current surge (may exceed 500 amps) followed by a stabilized reading. Crank for 10 seconds and record the stabilized reading _____ amps current draw.

   Task completed _____

10. Cranking circuit voltage drop tests. Use a DMM set to accurately measure 2 V-DC or less. Place one DMM test lead on the starter motor positive terminal and the other on the battery positive stud. Crank the engine and record the reading. This should typically be less than 0.5 V-DC but check OEM specification _____ V-DC cranking motor insulated circuit voltage drop. Place one DMM test lead on the starter ground stud and the other on the battery negative stud. Crank the engine and record the reading. Again this should typically be less than 0.5 V-DC but check OEM specification. Readings close to the maximum spec can indicate an imminent problem. _____ V-DC ground circuit voltage drop.

    Task completed _____

11. Cab insulated circuit voltage drop test. Locate the cab power supply positive terminal. This is usually located to the fuse/breaker box in the dash or on the power management module. Check OEM specifications and diagnostic routines. Use a DMM set to accurately measure 2 V-DC or less. Place one DMM test lead on the cab power supply positive terminal and the other on the battery positive stud. Turn the cab blower motor on full and all of the vehicle lights and accessories. The reading should typically be less than 0.2 V-DC but check OEM specifications. _____ V-DC cab insulated circuit voltage drop.

    Task completed _____

12. Cab ground circuit voltage drop test. Use a DMM set to accurately measure 2 V-DC or less. Place one DMM test lead on a known good ground on the dash and the other on the battery negative stud. Turn the cab blower motor on full and all of the vehicle lights and accessories. The reading should typically be less than 0.3 V-DC but check OEM specifications. _____ V-DC cab ground circuit voltage drop.

Task completed _____

STUDENT SELF-EVALUATION

Check	Level	Competency	Comments
	4	Mastered task	
	3	Competent but need further help	
	2	Needed a lot of help	
	0	Did not understand the task	

INSTRUCTOR EVALUATION

Check	Level	Competency	Comments
	4	Mastered task	
	3	Competent but needs further help	
	2	Requires more training	
	0	Unable to perform task	

ONLINE TASKS

Use a search engine to access the following Web pages and make a note of the vehicle lighting products they are marketing:

1. Grote
2. Delphi
3. Bosch

STUDY TIPS

Identify five key points in Chapter 10. Try to be as brief as possible.

Key point 1 _____

Key point 2 _____

Key point 3 _____

Key point 4 _____

Key point 5 _____

11 Diagnosis and Repair of Electronic Circuits

Objectives

After reading this chapter, you should be able to:

- Explain what is meant by sequential electronic troubleshooting.
- Perform tests on some key electronic components, including diodes and transistors.
- Define the acronym EST.
- Identify some types of EST in current use.
- Identify the levels of access and programming capabilities of each EST.
- Explain why electronic damage may be caused by electrostatic discharge and by using inappropriate circuit analysis tools.
- Describe the type of data that can be accessed by each EST.
- Identify the data may be read using the onboard flash codes and driver digital displays.
- Perform some basic electrical circuit diagnosis using a DMM.
- Identify the function codes on a typical DMM.
- Test some common input circuit components such as thermistors and potentiometers.
- Test semiconductor components such as diodes and transistors.
- Describe the full range of uses of handheld reader/programmer ESTs.
- Connect an EST to a vehicle data bus via the data connector and scroll through the display windows.
- Define the objectives of snapshot test.
- Describe how to use labscopes and PC-based oscilloscopes (Pico) to diagnose common truck electrical and electronic conditions.
- Perform electrical circuit analyses using Pico software.
- Outline the procedure required to use a PC and OEM software to read, diagnose, and reprogram vehicle electronic systems.
- Understand the importance of precisely completing each step when performing sequential troubleshooting testing of electronic circuits.
- Identify the SAE J1587/1939 codes for SAs, SPNs, PGNs, MIDs, PIDs, SIDs, and FMIs.
- Repair the sealed electrical connectors used in most electronic wiring harnesses.
- Interpret common ISO and DIN wiring symbols.
- Navigate generic, ISO, and DIN wiring schematics.
- Describe the operating principles and work safely around truck SRS and rollover protection systems.

PRACTICE QUESTIONS

1. What can be said of current flow if a circuit is described as open?
   a. excessive amps, may cause overheating
   b. high-resistance circuit, may cause overheating
   c. current is shorted to ground
   d. no current flows

2. What type of chassis data connector is required to access a SAE J1939 data bus?
   a. weatherproof
   b. weatherpac
   c. 6-pin Deutsch
   d. 9-pin Deutsch
3. When a DMM ammeter is used to measure current flow, which of the following should be true?
   a. All the current being measured passes through the DMM.
   b. It is connected in parallel with the circuit.
   c. The DMM fuse will probably blow.
   d. The reading produced is not going to be accurate.
4. Technician A says that a black J1939 chassis data bus connector may be used to connect an EST to a MY 2102 J1939 data bus. Technician B says that the reason for twisting the pair of wires in chassis data bus wire pair is to help insulate the bus from low-level radiation interference such as radio frequency (RF) waves. Who is correct?
   a. Technician A only
   b. Technician B only
   c. both A and B
   d. neither A nor B
5. What colors are used to identify the twisted wire pair used in a J1939 data bus?
   a. red and black
   b. yellow and green
   c. silver and blue
   d. brown and purple
6. Which GUI is found on most PCs in business and commercial use today?
   a. DOS
   b. windows
   c. leopard
   d. navigator
7. What diagnostic software is required to work on a multiplexed Freightliner truck?
   a. DOC
   b. ACOM
   c. toolbox
   d. ServiceLink
8. What is indicated when a CWS red light is illuminated?
   a. 3 seconds closing distance
   b. 2 seconds closing distance
   c. 1 second closing distance
   d. Collision has occurred.
9. What identifies the cathode on a diode?
   a. a dark band
   b. black terminal
   c. red terminal
   d. nothing
10. Take a look at the circuit schematic symbol for a standard SAE relay. What must happen before current can pass through the 87 terminal? Discuss.

JOB SHEET 11.1

Name _____ Station _____ Date _____

Test Potentiometer-Type Throttle Position Sensor (TPS) Resistance.

Performance Objective(s): Perform resistance tests on a potentiometer-type TPS using a DMM.

ASE Education Foundation Correlation

This job sheet addresses the following ASE Education Foundation task(s):

V. Electrical/Electronic Systems
   A. General
   1. Research vehicle service information, including vehicle service history, service precautions, and technical service bulletins. (IMMR, TST, MTST) P-1.
   2. Demonstrate knowledge of electrical/electronic series, parallel, and series-parallel circuits using principles of electricity (Ohm's Law). (IMMR, TST, MTST) P-1.
   3. Demonstrate proper use of test equipment when measuring source voltage, voltage drop (including grounds), current flow, continuity, and resistance. (IMMR, TST, MTST) P-1.
   4. Demonstrate knowledge of the causes and effects of shorts, grounds, opens, and resistance problems in electrical/electronic circuits. (IMMR) P-1.
   4. Demonstrate knowledge of the causes and effects of shorts, grounds, opens, and resistance problems in electrical/electronic circuits; identify and locate faults in electrical/electronic circuits. (TST, MTST) P-1.
   5. Use wiring diagrams to trace electrical/electronic circuits. (IMMR) P-1.
   5. Use wiring diagrams during the diagnosis (troubleshooting) of electrical/electronic circuit problems. (TST, MTST) P-1.
   9. Use appropriate electronic service tool(s) and procedures to check, record, and clear diagnostic codes; interpret digital multimeter (DMM) readings. (IMMR) P-2.
   9. Use appropriate electronic service tool(s) and procedures to diagnose problems; check, record, and clear diagnostic codes; interpret digital multimeter (DMM) readings. (TST, MTST) P-2.
   11. Identify electrical/electronic system components and configuration. (IMMR, TST, MTST) P-1.

   F. Instrument Cluster and Driver Information Systems
   2. Identify the sensor/sending units, gauges, switches, relays, bulbs/LEDs, wires, terminals, connectors, sockets, printed circuits, and control components/modules of the instrument cluster, driver information system, and warning systems. (IMMR) P-2.
   2. Identify faults in the sensor/sending units, gauges, switches, relays, bulbs/LEDs, wires, terminals, connectors, sockets, printed circuits, and control components/modules of the instrument cluster, driver information systems, and warning systems; determine needed action. (TST) P-2.
   2. Diagnose faults in the sensor/sending units, gauges, switches, relays, bulbs/LEDs, wires, terminals, connectors, sockets, printed circuits, and control components/modules of the instrument cluster, driver information systems, and warning systems; determine needed action. (MTST) P-2.

Tools and Materials: A MM, terminal extenders, a calculator, a truck with either an electronically managed engine or transmission, and the vehicle wiring schematic.

Protective Clothing: Standard shop apparel including coveralls or shop coat, safety glasses, and safety footwear.

PROCEDURE
Ensure that the shop LOTO is observed.

This test may be performed as a bench test or on the vehicle itself. Make sure that the truck is equipped with a potentiometer-type TPS that outputs an analog signal. Most trucks built before 2007 other than those with Caterpillar engines, will be OK to use for this test. Some trucks built after 2007 will use noncontact Hall effect TPS that will not be suitable for this test routine.

1. Separate the connector to the electronic foot pedal assembly (EFPA). Install terminal extenders on the TPS terminals. Use the DMM set in resistance test mode, and ground one of the DMM test probes. Use the vehicle wiring schematic to locate the signal terminal on the TPS, and place the other DMM test probe on that. Record the resistance value as R1.

   R1 _____

   Task completed _____

2. Now sweep the accelerator pedal through its full stroke, noting what happens to the resistance value measured by the DMM. The change in the resistance value reading should be smooth and even as the pedal angle increases. This test is best performed with an analog ohmmeter or a DMM with an analog bar scale.

   Smooth resistance sweep _____

   Record the upper and lower test values _____ / _____

   Task completed _____

3. At full pedal stroke, record the resistance reading as R2. Next, using the reference voltage (V-Ref) value (this is usually 5 V, but check the OEM service literature) and the two measured resistance values, use the calculator to calculate the current flow in each pedal location.

   Current flow at 0 pedal angle _____

   Current flow at full pedal travel _____

   Task completed _____

STUDENT SELF-EVALUATION

Check	Level	Competency	Comments
	4	Mastered task	
	3	Competent but need further help	
	2	Needed a lot of help	
	0	Did not understand the task	

INSTRUCTOR EVALUATION

Check	Level	Competency	Comments
	4	Mastered task	
	3	Competent but needs further help	
	2	Requires more training	
	0	Unable to perform task	

JOB SHEET 11.2

Name _____ Station _____ Date _____

Bench Test a Throttle Position Sensor (TPS) Signal Voltage.

Performance Objective(s): Perform a bench test to verify the output signal from a potentiometer-type TPS using a DMM.

ASE Education Foundation Correlation

This job sheet addresses the following ASE Education Foundation task(s):

V. Electrical/Electronic Systems
A. General
2. Demonstrate knowledge of electrical/electronic series, parallel, and series-parallel circuits using principles of electricity (Ohm's Law). (IMMR, TST, MTST) P-1.
3. Demonstrate proper use of test equipment when measuring source voltage, voltage drop (including grounds), current flow, continuity, and resistance. (IMMR, TST, MTST) P-1.
4. Demonstrate knowledge of the causes and effects of shorts, grounds, opens, and resistance problems in electrical/electronic circuits. (IMMR) P-1.
4. Demonstrate knowledge of the causes and effects of shorts, grounds, opens, and resistance problems in electrical/electronic circuits; identify and locate faults in electrical/electronic circuits. (TST, MTST) P-1.
11. Identify electrical/electronic system components and configuration. (IMMR, TST, MTST) P-1.

F. Instrument Cluster and Driver Information Systems
2. Identify the sensor/sending units, gauges, switches, relays, bulbs/LEDs, wires, terminals, connectors, sockets, printed circuits, and control components/modules of the instrument cluster, driver information system, and warning systems. (IMMR) P-2.
2. Identify faults in the sensor/sending units, gauges, switches, relays, bulbs/LEDs, wires, terminals, connectors, sockets, printed circuits, and control components/modules of the instrument cluster, driver information systems, and warning systems; determine needed action. (TST) P-2.
2. Diagnose faults in the sensor/sending units, gauges, switches, relays, bulbs/LEDs, wires, terminals, connectors, sockets, printed circuits, and control components/modules of the instrument cluster, driver information systems, and warning systems; determine needed action. (MTST) P-2.

Tools and Materials: A DMM, terminal extenders, a 5 V-DC power supply (any 110 V-AC to V-DC variable transformer) to the vehicle wiring schematic/test specifications appropriate for the TPS being tested.

Protective Clothing: Standard shop apparel including coveralls or shop coat, safety glasses, and safety footwear.

PROCEDURE

1. Using the vehicle wiring schematic, identify the three terminals on the TPS. Connect terminal extenders to the TPS terminals.

   Task completed _____

2. Connect the positive lead of the power supply unit to the TPS V-Ref terminal. Connect the negative lead of the power supply unit to the TPS ground terminal. Turn on the power.

   Task completed _____

3. Set the DMM to read V-DC with the scale set to read up to 5 V. Now place the negative lead from the DMM on the TPS ground terminal and the positive lead on the TPS signal terminal. Record the voltage value.

   Voltage at 0 pedal angle _____

   Task completed _____

4. Now sweep the pedal through its stroke while observing the changing voltage value. This test is best performed with an analog meter or a DMM with an analog bar scale. The change in voltage should correlate with pedal travel with no jumping or flat-spotting if the TPS is functioning properly.

   Task completed _____

5. Record the signal voltage value at full pedal travel. Now check the OEM specifications. Are the signal voltage readings within specifications?

   Voltage at full pedal travel _____

   Task completed _____

JOB SHEET 11.3

Name _____ Section _____ Date _____

Perform an Oscilloscope Cranking and Charging Circuit Test on a Truck Electrical System.

Performance Objectives: Use an oscilloscope to run a detailed analysis of a truck cranking and charging circuits and learn to interpret the waveform differences between normal and abnormal wave traces.

ASE Education Foundation Correlation

This job sheet addresses the following ASE Education Foundation task(s):

V. Electrical/Electronic Systems

A. General
1. Research vehicle service information, including vehicle service history, service precautions, and technical service bulletins. (IMMR, TST, MTST) P-1.
2. Demonstrate knowledge of electrical/electronic series, parallel, and series-parallel circuits using principles of electricity (Ohm's Law). (IMMR, TST, MTST) P-1.
3. Demonstrate proper use of test equipment when measuring source voltage, voltage drop (including grounds), current flow, continuity, and resistance. (IMMR, TST, MTST) P-1.
4. Demonstrate knowledge of the causes and effects of shorts, grounds, opens, and resistance problems in electrical/electronic circuits. (IMMR) P-1.
4. Demonstrate knowledge of the causes and effects of shorts, grounds, opens, and resistance problems in electrical/electronic circuits; identify and locate faults in electrical/electronic circuits. (TST, MTST) P-1.
5. Use wiring diagrams to trace electrical/electronic circuits. (IMMR) P-1.
5. Use wiring diagrams during the diagnosis (troubleshooting) of electrical/electronic circuit problems. (TST, MTST) P-1.

B. Battery System
1. Identify battery type and system configuration. (IMMR, TST, MTST) P-1.
2. Confirm proper battery capacity for application; perform battery state-of-charge test; perform battery capacity test, determine needed action. (IMMR, TST, MTST) P-1.

C. Starting System
1. Demonstrate understanding of starter system operation. (IMMR, TST, MTST) P-1.
2. Perform starter circuit cranking voltage and voltage drop tests. (IMMR) P-1.
2. Perform starter circuit cranking voltage and voltage drop tests; determine needed action. (TST, MTST) P-1.
3. Inspect starter control circuit switches, relays, connectors, terminals, wires, and harnesses (including over-crank protection). (IMMR) P-1.
3. Inspect and test starter control circuit switches (key switch, push button, and/or magnetic switch), relays, connectors, terminals, wires, and harnesses (including over-crank protection); determine needed action. (TST, MST) P-1.

D. Charging System
1. Identify and understand operation of the generator (alternator). (IMMR, TST, MTST) P-1.
5. Perform charging system voltage and amperage output tests; perform AC ripple test. (IMMR) P-1.
5. Perform charging system voltage and amperage output tests; perform AC ripple test; determine needed action. (TST, MTST).

Tools and Material: A scopemeter, oscilloscope, or Pico test kit. A truck with fully functional cranking and charging circuits. The test procedure outlined in this Job Sheet is based on using Pico software running in a Windows environment: access to Pico waveform libraries will help define the waveforms produced by the testing procedure.

Protective Clothing: Service facility PPE. Safety glasses should always be worn when working with chassis electrical systems.

PROCEDURE

Ensure the appropriate shop LOTO is used for the test procedure. This test assumes the use of Pico hardware and software but should also apply to most Scopemeters.

The battery, cranking, and charging circuit test is fast and accurate. First, ensure that the battery state of charge is satisfactory. Connect the Pico hardware to the truck circuit and establish communication between the Pico comm box and the computer you will be using. Figure 11.1 shows a Pico communication adapter with its A, B, C, and D connection points.

FIGURE 11-1 Pico communication adapter

TEST PROCEDURE

1. Connect a BNC cable from Channel A on the PicoScope communication adapter to the battery +, and to a reliable chassis ground point on the truck.

   Task completed _____

2. Next connect a 2000A current clamp to PicoScope comm box and place it on cranking motor positive cable. Figure 11.2 shows the connection circuit.

   Task completed _____

3. On the PC, run the PicoDiagnostics software setup wizard.

   Task completed _____

4. Select and run the cranking and charging diagnostics from the PicoScope menu.

   Task completed _____

5. Interpret the test result. This is displayed using a simple traffic light system: the test comprehensively covers all the starting and charging components.

   Task completed _____

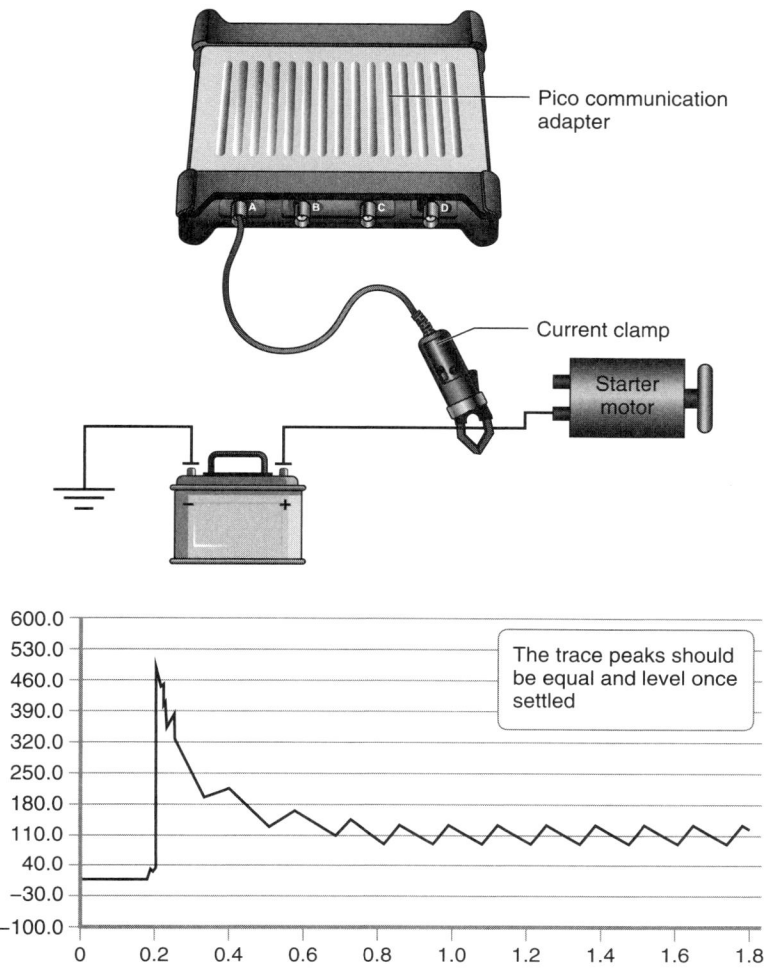

FIGURE 11-2 Pico charging and cranking circuit test

STUDENT SELF-EVALUATION

Check	Level	Competency	Comments

INSTRUCTOR EVALUATION

Check	Level	Competency	Comments

ONLINE TASKS

Use a search engine to access the following Web pages and make a list of any emerging technologies used to read and diagnose vehicle electronic systems:

1. MPSI
2. Fluke
3. Delphi Lucas

STUDY TIPS

Identify 5 key points in Chapter 11. Try to be as brief as possible.

Key point 1 _____

Key point 2 _____

Key point 3 _____

Key point 4 _____

Key point 5 _____

12 Multiplexing

Objectives

After reading this chapter, you should be able to:

- Describe a typical vehicle data bus.
- List the key data bus hardware components.
- Define the word *multiplexing*.
- Describe how multiplexing can make data exchange more efficient.
- Outline how a J1939/CAN 2.0 data bus functions.
- Define the terms "source address" and "suspect parameter number."
- Access J1587/1708 and J1939 data buses using a data connector.
- Explain how a "smart" ladder switch operates.
- Identify the fields that make up a data frame message packet on a truck data bus transaction.
- Explain how messages are prioritized on a serial data bus.
- Outline some of the 2014 changes to J1939.
- Identify the different generations of J1939 data bus connectors.
- Explain how FETs are used as relays to effect data bus outcomes.
- Access a controller source address (SA) on a truck chassis data bus with multiple networked electronic systems.
- Describe the composition of a diagnostic trouble code.
- Outline the procedure required to access a failure mode indicator (FMI) using electronic service tools.
- Identify the bus topology used in semi- and fully- autonomous trucks.
- Describe the scope characteristics of a J1939 message packet.

PRACTICE QUESTIONS

1. When two or more truck computerized systems such as the engine and transmission electronics are capable of sharing data at high speeds on a communications bus, this is usually known as:
   a. data processing
   b. data filtration
   c. multiplexing
   d. handshaking

2. How many bits are there in a byte?
   a. 2
   b. 4
   c. 8
   d. 256

3. Which of the following is true of a green J1939 chassis data bus connector?
   a. It has 9 pins.
   b. It is backward compatible to pre-2014 J1939 buses.
   c. It cannot access post-2014 Volvo-Mack J1939.
   d. all of the above

4. Which of the following would be typical "wake-up" source(s) for a typical truck multiplexed data bus?
   a. headlight switch
   b. brake light switch
   c. keyless door lock
   d. all of the above

5. Which of the following would be most likely to occur if you were to physically disconnect a "smart" switch on a truck dash?
   a. log a failure code
   b. shut down the data bus
   c. disable the vehicle
   d. disable the ability to read any MIDs networked to the data bus

6. When examining how a FET functions, which of the following best describes the role of the gate?
   a. acts as a resistor
   b. controls resistance through the device
   c. inputs system voltage (V-Bat)
   d. acts as a diode

7. What is another way of saying *algorithm*?
   a. networking
   b. mapping
   c. switching
   d. downloading

8. What is the term used to describe a communication signal that is converted to a radio frequency signal and then superimposed over a trailer cord, 12-volt auxiliary power wire?
   a. power-line carrier
   b. feedback
   c. wireless networking
   d. microwave networking

9. When the resistance status on a multiplex smart switch is altered, what happens immediately after?
   a. A hard wire electrical signal is generated.
   b. A "packet" is broadcast to the data bus.
   c. An FET is grounded.
   d. System voltage actuates the switched circuit.

10. When repairing a low bus twisted wire pair, which of the following steps must be observed?
    a. Maintain the existing twisting cycles.
    b. Maintain the correct gauge size.
    c. Use the correct solder.
    d. all of the above

JOB SHEET 12.1

Name _____ Station _____ Date _____

Electronically Identify the Controllers on a Truck Chassis and Identify Each by SA/MID.

Performance Objective(s): Use a data link connection to a truck chassis and EST to identify the controllers networked on the data bus and note any DTCs and FMIs logged.

ASE Education Foundation Correlation

This job sheet addresses the following ASE Education Foundation task(s):

V. Electrical/Electronic Systems
  A. General
  1. Research vehicle service information, including vehicle service history, service precautions, and technical service bulletins. (IMMR, TST, MTST) P-1.
  5. Use wiring diagrams to trace electrical/electronic circuits. (IMMR) P-1.
  5. Use wiring diagrams during the diagnosis (troubleshooting) of electrical/electronic circuit problems. (TST, MTST) P-1.
  9. Use appropriate electronic service tool(s) and procedures to check, record, and clear diagnostic codes; interpret digital multimeter (DMM) readings. (IMMR) P-2.
  9. Use appropriate electronic service tool(s) and procedures to diagnose problems; check, record, and clear diagnostic codes; interpret digital multimeter (DMM) readings. (TST, MTST) P-2.
  10. Check for malfunctions caused by faults in the data bus communications network. (IMMR, TST) P-2.
  10. Diagnose faults in the data bus communications network; determine needed action. (MTST) P-2.
  11. Identify electrical/electronic system components and configuration. (IMMR, TST, MTST) P-1.

  F. Instrument Cluster and Driver Information Systems
  2. Identify the sensor/sending units, gauges, switches, relays, bulbs/LEDs, wires, terminals, connectors, sockets, printed circuits, and control components/modules of the instrument cluster, driver information system, and warning systems. (IMMR) P-2.
  2. Identify faults in the sensor/sending units, gauges, switches, relays, bulbs/LEDs, wires, terminals, connectors, sockets, printed circuits, and control components/modules of the instrument cluster, driver information systems, and warning systems; determine needed action. (TST) P-2.
  2. Diagnose faults in the sensor/sending units, gauges, switches, relays, bulbs/LEDs, wires, terminals, connectors, sockets, printed circuits, and control components/modules of the instrument cluster, driver information systems, and warning systems; determine needed action. (MTST) P-2.

Tools and Materials: An EST with appropriate software and chassis data bus connection hardware, along with a truck equipped with data bus communications.

Protective Clothing: Standard shop apparel including coveralls or shop coat, safety glasses, and safety footwear.

PROCEDURE
Ensure that the shop LOTO is observed.

Connect the EST to the chassis data bus and scroll the SAs/MIDs observed making a note of any DTCs and FMIs logged in Table 12-1. The language used to describe the controllers in this table use International Navistar preferred terms but substitute other acronyms if using a different OEM chassis.

Type of data connector used (circle):

J1708 (6-pin)   J1939 (9-pin black)   J1939 (9-pin green)   J1962 (16-pin)

Bus communication channel (circle):

J1587/1708   J1939

Table 12-1

Module Name	Source Address	DTCs/FMIs logged
Engine Control Module (ECM)	00	
Transmission Control Module (TCM)	03	
Shift Selector Module	05	
Antilock Brake System (ABS)	11	
Electronic Gauge Cluster (EGC)	23	
Compass Module	28	
Body Controller	33	
Vehicle Sensor Module (VSM)	39	
Vehicle Information Display (VID)	40	
Tire Pressure Monitoring System (TPMS)	51	
Rear HVAC	58	
Aftertreatment Module	61	
Telematics Module	74	
Auxiliary Gauge Switch Pack (AGSP)	132	
Hybrid Electric Vehicle (HEV)	239	

STUDENT SELF-EVALUATION

Check	Level	Competency	Comments
	4	Mastered task	
	3	Competent but need further help	
	2	Needed a lot of help	
	0	Did not understand the task	

INSTRUCTOR EVALUATION

Check	Level	Competency	Comments
	4	Mastered task	
	3	Competent but needs further help	
	2	Requires more training	
	0	Unable to perform task	

JOB SHEET 12.2

Name _____ Station _____ Date _____

Identify the Physical Location of Controllers on a Truck Chassis and Identify Each by SA/MID.

Performance Objective(s): Use a schematic to identify the physical location of each controller networked to the chassis data bus on a truck.

ASE Education Foundation Correlation

This job sheet addresses the following ASE Education Foundation task(s):

V. Electrical/Electronic Systems
  A. General
   1. Research vehicle service information, including vehicle service history, service precautions, and technical service bulletins. (IMMR, TST, MTST) P-1.
   10. Check for malfunctions caused by faults in the data bus communications network. (IMMR, TST) P-2.
   10. Diagnose faults in the data bus communications network; determine needed action. (MTST) P-2.
   11. Identify electrical/electronic system components and configuration. (IMMR, TST, MTST) P-1.

Tools and Materials: A chassis schematic such as that shown in Figure 12–1 and a late-model truck equipped with a J1939 data bus.

Protective Clothing: Standard shop apparel including coveralls or shop coat, safety glasses, and safety footwear.

PROCEDURE
Ensure that the shop LOTO is observed.

Using Figure 12–1 (this shows a medium duty truck chassis), identify the location of each SA/MID controller networked to the chassis data bus by filling in the blank boxes. Add more if required. You can cross check this with the figure in Chapter 12 of the textbook but bear in mind that each OEM uses different terms to describe the controllers used on their chassis.

Type of data connector used (circle):

J1708 (6-pin)   J1939 (9-pin black)   J1939 (9-pin green)   J1962 (16-pin)

Bus communication channel (circle):

J1587/1708   J1939

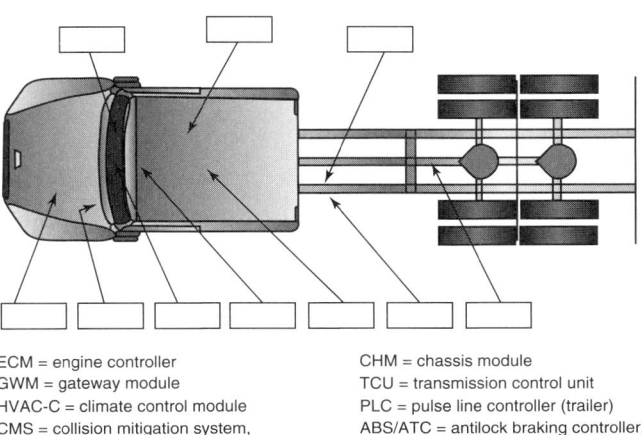

ECM = engine controller
GWM = gateway module
HVAC-C = climate control module
CMS = collision mitigation system, wingman radar module
ICU = instrument cluster unit
CHM = chassis module
TCU = transmission control unit
PLC = pulse line controller (trailer)
ABS/ATC = antilock braking controller, automatic traction control
TM = telematics module

FIGURE 12–1 In the appropriate box, identify each SA/MID location on a truck chassis: OEMs may use different acronyms for controllers than those here, but attempt to cross reference them.

Multiplexing 181

STUDENT SELF-EVALUATION

Check	Level	Competency	Comments
	4	Mastered task	
	3	Competent but need further help	
	2	Needed a lot of help	
	0	Did not understand the task	

INSTRUCTOR EVALUATION

Check	Level	Competency	Comments
	4	Mastered task	
	3	Competent but needs further help	
	2	Requires more training	
	0	Unable to perform task	

JOB SHEET 12.3

Name _____ Station _____ Date _____

Diagnose, Locate, and Remedy a DTC on Truck Data Bus Using the Appropriate Diagnostic Hardware and Software.

Performance Objective(s): Use OEM electronic diagnostic routines to locate a simple DTC and FMI, make a repair, and erase the logged codes.

ASE Education Foundation Correlation

This job sheet addresses the following ASE Education Foundation task(s):

V. Electrical/Electronic Systems
  A. General
  1. Research vehicle service information, including vehicle service history, service precautions, and technical service bulletins. (IMMR, TST, MTST) P-1.
  9. Use appropriate electronic service tool(s) and procedures to check, record, and clear diagnostic codes; interpret digital multimeter (DMM) readings. (IMMR) P-2.
  9. Use appropriate electronic service tool(s) and procedures to diagnose problems; check, record, and clear diagnostic codes; interpret digital multimeter (DMM) readings. (TST, MTST) P-2.
  10. Check for malfunctions caused by faults in the data bus communications network. (IMMR, TST) P-2.
  10. Diagnose faults in the data bus communications network; determine needed action. (MTST) P-2.
  11. Identify electrical/electronic system components and configuration. (IMMR, TST, MTST) P-1.

Tools and Materials: A late-model truck equipped with a J1939 data bus and an appropriate EST and data connection hardware.

Protective Clothing: Standard shop apparel including coveralls or shop coat, safety glasses, and safety footwear.

PROCEDURE
Ensure that the shop LOTO is observed.

Work in pairs: take turns in disconnecting a single, easy to access sensor on the truck that will result in logging a DTC, challenging the other to locate and remedy the problem created. Examples of sensors to target are:

- wheel speed sensor
- transmission tailshaft speed sensor
- ambient temperature sensor

Connect the EST to the data bus.

Type of data connector used (circle):

J1708 (6-pin)   J1939 (9-pin black)   J1939 (9-pin green)   J1962 (16-pin)

Bus communication channel (circle):

J1587/1708   J1939

1. Sensor disconnected _____
2. SA/MID circuit _____
3. DTC _____
4. FMI _____
5. Reconnect sensor _____
6. Clear codes _____

STUDENT SELF-EVALUATION

Check	Level	Competency	Comments
	4	Mastered task	
	3	Competent but need further help	
	2	Needed a lot of help	
	0	Did not understand the task	

INSTRUCTOR EVALUATION

Check	Level	Competency	Comments
	4	Mastered task	
	3	Competent but needs further help	
	2	Requires more training	
	0	Unable to perform task	

ONLINE TASKS

Use a search engine and input the key words listed next. You should be able to access more recent information than that in the textbook. See if you can identify any emerging trends.

1. Multiplexing
2. FET
3. CAN 2.0
4. Smart switches
5. Ladder logic

STUDY TIPS

Identify 5 key points in Chapter 12. Try to be as brief as possible.

Key point 1 _____

Key point 2 _____

Key point 3 _____

Key point 4 _____

Key point 5 _____

13 Hydraulics

Objectives

After reading this chapter, you should be able to:
- Explain fundamental hydraulic principles.
- Apply the laws of hydraulics.
- Calculate force, pressure, and area.
- Describe the function of pumps, valves, actuators, and motors.
- Describe the construction of hydraulic conductors and couplers.
- Outline the properties of hydraulic fluids.
- Identify graphic symbols.
- Interpret a hydraulic schematic.
- Perform maintenance procedures on truck hydraulic systems.
- Identify safe practices when working with mobile hydraulics.
- Safely troubleshoot hydraulic leaks and understand the danger of hydraulic pinhole injection injuries and how to respond if you suspect such an injury.

PRACTICE QUESTIONS

1. In which of the following applications would you be more likely to find a wet line kit?
   a. van-type trailer
   b. dump truck
   c. linehaul tractor
   d. school bus

2. Which of the following substances is NOT a fluid at ambient summer temperatures?
   a. water
   b. oxygen
   c. oil
   d. salt

3. The science of moving liquids to transmit energy such as the principle used in a torque converter is known as:
   a. hydrodynamics
   b. hydrostatics
   c. fluid power
   d. kinetics

4. Who first said these words: *Pressure applied to a confined liquid is transmitted in all directions with equal force*?
   a. Bernoulli
   b. Charles
   c. Einstein
   d. Pascal

5. Which of the following values is exactly equivalent to an atmospheric pressure value of 14.7 psi?
   a. 101.3 kPa
   b. 1 bar
   c. 100 atms
   d. 1 MPa

6. Technician A says that 29.9 inches (760 mm) of Hg (mercury) is approximately equivalent to 1 bar (metric unit of atmosphere). Technician B says that gauge pressure has its zero point when subjected to atmospheric pressure at sea level. Who is correct?
   a. Technician A only
   b. Technician B only
   c. both A and B
   d. neither A nor B

7. Technician A says that any single-acting hydraulic cylinder can exert force only on an outward stroke. Technician B says that double-acting hydraulic cylinders can exert force either inward or outward. Who is correct?
   a. Technician A only
   b. Technician B only
   c. both A and B
   d. neither A nor B

8. What is a good example of an open-center, truck hydraulic circuit?
   a. power steering
   b. lift gate circuit
   c. COE cab lift system
   d. torque converter

9. What is the usual method of rating truck hydraulic pump performance capability?
   a. flow rate (gpm) at a specified driven rpm
   b. peak pressure output
   c. pressure output at rated speed
   d. flow rate (gpm) at peak pressure

10. Which of the following hydraulic oils has an approximate flow rating to ISO 32 rated oil?
    a. SAE 5-grade
    b. SAE 10-grade
    c. SAE 32-grade
    d. SAE 46-grade

JOB SHEET 13.1

Name _____ Station _____ Date _____

Test the Efficiency of a Hydraulic Pump.

Performance Objective(s): Evaluate the efficiency of a hydraulic pump on a truck hydraulic system using a hydraulic circuit analyzer.

ASE Educational Foundation Correlation

This job sheet addresses the following ASE Educational Foundation task(s):

VIII. Hydraulics
  A. General
   1. Research vehicle service information, including vehicle service history, service precautions, fluid type, and technical service bulletins. (IMMR, TST, MTST) P-3.
   3. Identify hydraulic system components; locate filtration system components; service filters and breathers. (IMMR, TST, MTST) P-3.
   4. Check fluid level and condition; purge and/or bleed system; take a hydraulic fluid sample for analysis; determine needed action. (IMMR, TST, MTST) P-3.
   8. Perform system temperature, pressure, flow, and cycle time tests; determine needed action. (MTST) P-3.
  B. Pumps
   2. Determine pump type, rotation, and drive system. (MTST) P-3.
   4. Inspect pump inlet and outlet for restrictions and leaks; determine needed action. (MTST) P-3.
  C. Filtration/Reservoirs (Tanks)
   1. Identify type of filtration system; verify filter application and flow direction. (MTST) P-3.
  D. Hoses, Fittings, and Connections
   2. Inspect hoses and connections for leaks, proper routing, and proper protection; determine needed action. (MTST) P-3.
   3. Assemble hoses, tubes, connectors, and fittings. (MTST) P-3.
  E. Control Valves
   1. Pressure test system safety relief valve; determine needed action. (MTST) P-3.
   2. Perform control valve operation pressure and flow tests; determine needed action. (MTST) P-3.

Tools and Materials: A hydraulic circuit analyzer (flow meter, pressure gauge, flow volume valve, and thermometer) and a truck with a functional hydraulic system

Protective Clothing: Standard shop apparel including coveralls or shop coat, safety glasses, and safety footwear

PROCEDURE
Ensure that the shop LOTO is observed.

1. Relieve any residual pressure in the hydraulic circuit to be tested, then Tee into the circuit with the hydraulic circuit analyzer.

   Task completed _____

2. Depending on how the hydraulic pump is driven (direct-engine, PTO, etc.), run the engine at the speed specified for testing the hydraulic circuit. Usually this will be around 1,500 rpm.

   Task completed _____

3. Observe and record the system pressures.

   Peak system pressure _____ psi

   Flow volume _____ gpm

   Task completed _____

Hydraulics 187

4. Close down the flow volume valve until a pressure of 2,000 psi (or the OEM-specified test value) is achieved. Record the actual pressure value and flow volume in gpm.

   Peak system pressure value _____

   Flow volume _____ gpm

   Task completed _____

5. Now divide the gpm value recorded in step 3 by the gpm value recorded in step 4. If the resulting flow reduction registers within 10 percent or less, the pump can be said to be operating at 100 percent efficiency. For instance, if the reading is 100 gpm in step 3 and 91 gpm in step 4, pump efficiency is 91 percent: Because this is within the 10 percent specification, pump efficiency can be said to be 100 percent. However, if the value drops out of the 10 percent window, express as calculated pump efficiency. Example:

   Test 3 _____ 100 gpm

   Test 4 _____ 78 gpm = 78 percent efficient

   Task completed _____

STUDENT SELF-EVALUATION

Check	Level	Competency	Comments
	4	Mastered task	
	3	Competent but need further help	
	2	Needed a lot of help	
	0	Did not understand the task	

INSTRUCTOR EVALUATION

Check	Level	Competency	Comments
	4	Mastered task	
	3	Competent but needs further help	
	2	Requires more training	
	0	Unable to perform task	

ONLINE TASKS

Use a search engine and access the OEMs listed next. Make a list of the products each markets, and then comment on which companies do the best job of presenting their products.

1. Vickers/Eaton Hydraulics
2. Gates
3. Muncie Power

STUDY TIPS

Identify 5 key points in Chapter 13. Try to be as brief as possible.

Key point 1 _____

Key point 2 _____

Key point 3 _____

Key point 4 _____

Key point 5 _____

14 Clutches

Objectives

After reading this chapter, you should be able to:

- Outline the operating principles of a clutch.
- Identify the components of a clutch assembly.
- Explain the differences between centrifugal, pull-type, and push-type clutches.
- Describe the procedure for adjusting manual and self-adjusting clutches.
- Explain how to properly adjust the external linkage of a clutch.
- Describe the function of a clutch brake.
- Outline the procedures required to work safely around heavy-duty clutches.
- Describe how an ECU-managed clutch functions.
- Troubleshoot a clutch for wear and damage.
- Identify some typical clutch defects and explain how to repair them.
- Outline the procedure for removing and replacing a clutch.
- Explain the difference between a three-pedal and a two-pedal clutch system.

PRACTICE QUESTIONS

1. When installing a 15½-inch clutch pack, Technician A uses guide studs installed in the flywheel to support the weight. Technician B uses a splined pilot shaft when installing both 14- and 15½-inch clutch packs. Who is correct?
   a. Technician A only
   b. Technician B only
   c. both A and B
   d. neither A nor B

2. Technician A says that most vocational trucks use rigid clutch friction discs. Technician B says that rigid clutch friction discs are less able to absorb torsional shock. Who is correct?
   a. Technician A only
   b. Technician B only
   c. both A and B
   d. neither A nor B

3. Which of the following is a clutch driving member?
   a. pilot bearing
   b. throwout bearing
   c. transmission input shaft
   d. pressure plate

4. Which of the following dimensions is closest to being correct for clutch pedal free play measured at the pedal?
   a. ½ inch (12 mm)
   b. 1½ inches (38 mm)
   c. 2 inches (50 mm)
   d. 2½ inches (62 mm)

5. Which of the following dimensions is closest to being correct for clutch brake squeeze?
   a. 1 inch (25 mm)
   b. 1½ inches (38 mm)
   c. 2 inches (50 mm)
   d. 2½ inches (62 mm)
6. Which of the following could cause clutch hang-up?
   a. seized pilot bearing
   b. cocked intermediate drive plate pins
   c. seized throwout bearing
   d. stretched clutch linkage rod
7. Which of the following driving techniques can cause glazed clutch friction faces?
   a. starting off in high gear
   b. riding the clutch pedal
   c. holding the truck on an incline with the clutch pedal
   d. all of the above
8. What is the function of coaxial springs in a clutch friction plate?
   a. dampens driveline torsionals
   b. dampens engine torsionals
   c. reduces clutch chatter
   d. all of the above
9. Which of the following is more likely to cause the spline tangs to shear in a clutch brake?
   a. clutch brake squeeze below specification
   b. excessive free pedal setting
   c. attempting to brake the truck with the clutch brake
   d. riding the clutch pedal
10. If a pot-type clutch has the flywheel face machined, what else must also be machined?
    a. flywheel pot flange
    b. pilot bearing bore
    c. intermediate plate drive lugs
    d. ring gear

JOB SHEET 14.1

Name _____ Station _____ Date _____

Adjust a Heavy-Duty, Pull-Type Clutch.

Performance Objective(s): Learn the procedure for adjusting a heavy-duty, pull-type clutch on a functional truck chassis.

ASE Education Foundation Correlation

This job sheet addresses the following ASE Education Foundation task(s):

II. Drive Train
  B. Clutch
   1. Inspect and adjust clutch, clutch brake, linkage, cables, levers, brackets, bushings, pivots, springs, and clutch safety switch (includes push-type and pull-type); check pedal height and travel; determine needed action. (IMMR, TST, MTST) P-1.
   4. Inspect, adjust, lubricate, or replace release (throw-out) bearing, sleeve, bushings, springs, housing, levers, release fork, fork pads, rollers, shafts, and seals. (TST, MTST) P-1.
   6. Inspect, adjust, and/or replace two-plate clutch pressure plate, clutch discs, intermediate plate, and drive pins/lugs. (TST, MTST) P-1.
   7. Inspect and/or replace clutch brake assembly; inspect input shaft and bearing retainer; determine needed action. (TST, MTST) P-1.
   8. Inspect, adjust, and/or replace self-adjusting/continuous-adjusting clutch mechanisms. (TST, MTST) P-1.

Tools and Materials: A truck with a fully functional drivetrain, standard shop hand tools, a clutch adjusting ring tool, and standard measuring tools.

Protective Clothing: Standard shop apparel including coveralls or shop coat, safety glasses, and safety footwear.

PROCEDURE
Ensure that the shop LOTO is observed.

The procedure outlined here is that used to adjust a clutch with no automatic adjust mechanism. The procedure for adjusting a clutch with auto adjust is similar, but if an auto adjust clutch requires adjustment there has usually been a malfunction of the adjusting mechanism, which should be investigated.

1. Drive the truck into the shop, put the brake system into park, place the transmission into neutral, and block the wheels.

   Task completed _____

2. Check the clutch free play (free pedal) and clutch brake squeeze at the clutch pedal. This determines the need for adjustment. Typically, clutch free play should be around 1–1½ inches (25–37 mm) and clutch brake squeeze around 1 inch (25 mm) measured at the pedal. This can vary by chassis and application, so check the OEM specifications.

   Task completed _____

3. Assume that the clutch requires adjustment. Never adjust the external linkage unless the clutch brake squeeze is out of specification. Figure 14–1 and Figure 14–2 show two types of external clutch linkage.

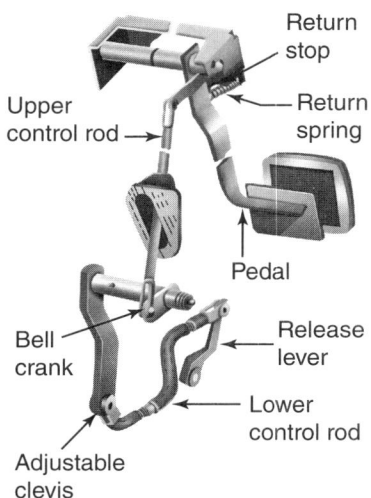

FIGURE 14-1 Adjustment to this external linkage is made by changing the position of the clevis on the end of the lower control arm.

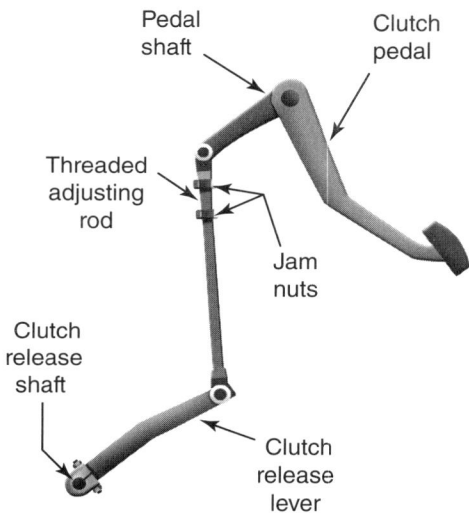

FIGURE 14-2 The linkage on this conventional truck is adjusted by lengthening or shortening the control rod.

If clutch brake squeeze is out of specification, it is an indication that either the release linkage has been tampered with or there is a problem with the clutch brake or release yoke assembly. Remove the bell housing inspection plate. Rotate the flywheel either by using a manual ratchet and gear barring tool or by bunting the starter so that the clutch ring lock plate is located at BDC. Remove the bolt that holds the lock plate in place between the clutch ring adjustment lugs, and then remove the lock plate.

Task completed _____

4. Install the clutch ring adjusting tool, inserting the clutch ring adjusting tool bolt in the lock plate fastener location.

Task completed _____

5. With someone in the cab releasing the clutch (pedal to the floor), use the clutch adjusting tool to rotate the clutch adjust ring clockwise (CW). This will increase clutch free play, so have the person in the cab check free play after every couple of lugs are rotated. When the free play dimension is correct at the pedal, the clutch should be adjusted properly.

Task completed _____

6. Next check the free play dimension between the release (throwout) bearing and the release yoke fingers. It should be 0.125 inch (3 mm) as shown in Figure 14–3.

Task completed _____

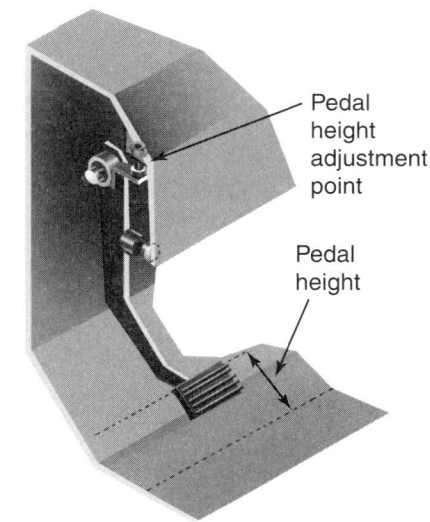

FIGURE 14–3 When properly adjusted, there should be ⅛ inch clearance between the release fork and the boss on the bearing, producing 1.5 inch (37 mm) free pedal.

7. Depress the clutch pedal and install the lock plate. This is important because the lock plate may not be perfectly aligned between the adjustment lugs and some rotation of the clutch ring may be required. Replace the clutch inspection plate.

Task completed _____

8. Test drive the truck, checking the performance of the clutch, looking for clutch slippage, chatter, vibration, and hang-up.

Task completed _____

STUDENT SELF-EVALUATION

Check	Level	Competency	Comments
	4	Mastered task	
	3	Competent but need further help	
	2	Needed a lot of help	
	0	Did not understand the task	

INSTRUCTOR EVALUATION

Check	Level	Competency	Comments
	4	Mastered task	
	3	Competent but needs further help	
	2	Requires more training	
	0	Unable to perform task	

JOB SHEET 14.2

Name _____ Station _____ Date _____

Install a 15½-Inch Clutch Assembly.

Performance Objective(s): Install a 15½-inch clutch pack with a 2-inch clutch shaft.

ASE Education Foundation Correlation

This job sheet addresses the following ASE Education Foundation task(s):

II. **Drive Train**
   B. **Clutch**
   1. Inspect and adjust clutch, clutch brake, linkage, cables, levers, brackets, bushings, pivots, springs, and clutch safety switch (includes push-type and pull-type); check pedal height and travel; determine needed action. (IMMR, TST, MTST) P-1.
   2. Inspect clutch master cylinder fluid level; check clutch master cylinder, slave cylinder, lines, and hoses for leaks and damage; determine needed action. (IMMR, TST, MTST) P-1.
   3. Inspect, adjust, repair, and/or replace hydraulic clutch slave and master cylinders, lines, and hoses; bleed system. (TST, MTST) P-2.
   4. Inspect, adjust, lubricate, or replace release (throw-out) bearing, sleeve, bushings, springs, housing, levers, release fork, fork pads, rollers, shafts, and seals. (TST, MTST) P-1.
   5. Inspect, adjust, and/or replace single-disc clutch pressure plate and clutch disc. (TST, MTST) P-1.
   6. Inspect, adjust, and/or replace two-plate clutch pressure plate, clutch discs, intermediate plate, and drive pins/lugs. (TST, MTST) P-1.
   7. Inspect and/or replace clutch brake assembly; inspect input shaft and bearing retainer; determine needed action. (TST, MTST) P-1.
   8. Inspect, adjust, and/or replace self-adjusting/continuous-adjusting clutch mechanisms. (TST, MTST) P-1.
   9. Inspect and/or replace pilot bearing. (TST, MTST) P-1.

Tools and Materials: A heavy-duty truck with an SAE 15½-inch flywheel, a 2-inch pilot shaft (a used clutch shaft with the drive gear machined flush is OK), a bearing driver (if the pilot bearing is to be replaced), and either a hoist or a mobile clutch jack.

Protective Clothing: Standard shop apparel including coveralls or shop coat, safety glasses, and safety footwear.

PROCEDURE
Ensure that the shop LOTO is observed.

1. Check the pilot bearing in the flywheel. If this requires replacement, pull with a bearing puller and drive the new one in with a bearing driver.

   Task completed _____

2. Check the intermediate drive plate lugs for nicks, burrs, and wear. Ensure that a new intermediate drive plate has all the sharp edges at the drive recesses dressed with a file. Replace the drive lugs in the flywheel if necessary. These are retained by recessed set screws. Remove and check that the new drive lugs allow the intermediate plate to slide over them without hanging up. Use the file to ensure that sharp edges on either the intermediate plate recesses or the drive lugs do not cause binding.

   Task completed _____

3. Check the orientation of the two friction discs. These are marked "flywheel side" (inboard disc) and "transmission" (outboard disc), so ensure that each faces in the correct direction. Install the transmission side friction disc onto the clutch pilot shaft first. Next place the flywheel side (inboard) disc into the flywheel pot and slide the intermediate plate over the flywheel pot drive lugs. Find the inboard disc splines with clutch pilot shaft (which already has the outboard disc mounted on it) and lift both friction discs so that the pilot bore of the pilot shaft can be inserted into the flywheel pilot bearing. Some pressure will have to be applied onto the pilot shaft to ensure that the friction discs and intermediate plates stay in place.

   Task completed _____

4. Now raise the clutch pressure plate assembly into position to slide over the pilot shaft. Use either a hydraulic clutch jack or hoist from above. Which of these you use will depend on access—it is inadvisable to attempt to lift a

15½-inch clutch, especially in a restricted access chassis, due to the weight of the assembly. A new clutch pack will have the pressure plate springs compressed by chocks of wood—this facilitates clutch installation. Pass the assembly through the pilot shaft and onto at least two guide studs installed in the flywheel.

Task completed _____

5. Install all the clutch-to-flywheel fasteners finger-tight. Next, evenly torque down the fasteners. This should cause the wooden chocks between the pressure plate and the release bearing to drop out. They can be disposed of.

Task completed _____

6. Remove the pilot shaft. If the shaft is difficult to remove, use a nylon hammer to tap it from side to side while pulling outboard. After the pilot shaft has been removed, use a flashlight to visually check the alignment of the friction disc splines and the pilot bearing

Task completed _____

STUDENT SELF-EVALUATION

Check	Level	Competency	Comments
	4	Mastered task	
	3	Competent but need further help	
	2	Needed a lot of help	
	0	Did not understand the task	

INSTRUCTOR EVALUATION

Check	Level	Competency	Comments
	4	Mastered task	
	3	Competent but needs further help	
	2	Requires more training	
	0	Unable to perform task	

ONLINE TASKS

Use a search engine and input the clutch OEMs and rebuilders listed next. Make a note of their products and specialties. Obtain a clutch pack serial number and compare the cost of a new clutch versus a rebuilt clutch.

1. Eaton Fuller
2. Spicer Dana Corporation
3. Haldex

STUDY TIPS

Identify 5 key points in Chapter 14. Try to be as brief as possible.

Key point 1 _____

Key point 2 _____

Key point 3 _____

Key point 4 _____

Key point 5 _____

15 Standard Transmissions

Objectives

After reading this chapter, you should be able to:
- Define the function of a transmission in the driveline of a truck.
- Identify the types of gears used in truck transmissions.
- Interpret the language used to describe gear trains and calculate gear pitch and gear ratios.
- Explain the relationship between speed and torque from input to output in different gear arrangements.
- Identify the major components in a typical transmission—including input and output shafts, main shaft and countershaft gears, and shift mechanisms.
- Describe the shift mechanisms used in heavy-duty manual truck transmissions.
- Outline the role of main and auxiliary (compound) gear sections in a typical transmission, and trace the power-flow from input to output in different ratios.
- Describe the operating principles of range shift and splitter shift air systems.
- Define the roles of transfer cases and PTOs in heavy-duty truck operation.

PRACTICE QUESTIONS

1. Which of the following gear ratios would reduce the output speed of a transmission the most?
   a. 3.60:1
   b. 2.85:1
   c. 1.20:1
   d. 0.73:1

2. Which OEM would be associated with triple countershaft transmissions?
   a. Freightliner
   b. International Navistar
   c. Mack Trucks
   d. Kenworth Trucks

3. Which of the following best describes the role of an auxiliary section in a truck standard transmission?
   a. multiplies torque
   b. multiplies available ratios
   c. reduces transmission length
   d. enables faster road speeds

4. How many countershafts are there in the forward section of a Roadranger transmission?
   a. one
   b. two
   c. three
   d. four

5. Which of the following best explains the need to have backlash between meshing gears?
   a. to permit bottoming clearance
   b. to prevent climbing
   c. to allow for heat expansion and lubrication of gears
   d. to limit coasting whine

6. What term describes the gear surface area closest to the outside diameter?
   a. pitch diameter
   b. addendum
   c. root diameter
   d. dedendum
7. If a drive gear is required to rotate a driven gear in the same direction, which of the following is a requirement?
   a. drive gear pitch diameter larger than driven gear
   b. driven gear must have twice the number of teeth as the drive gear
   c. idler gear
   d. helical gear
8. If a transmission is described as nonsynchronized, which of the following is a required driving technique?
   a. double clutching
   b. a good ear
   c. feathering the clutch
   d. rapid gear engagements
9. Which of the following is required to activate the low-low ratio in a deep-reduction transmission?
   a. actuate split shifter
   b. actuate dash deep reduction valve
   c. stop the truck
   d. all of the above
10. Which of the following terms is often used to describe a type of transfer case?
    a. compound
    b. PTO
    c. shift tower
    d. dropbox

JOB SHEET 15.1

Name _____ Station _____ Date _____

Remove a Truck Standard Transmission.

Performance Objective(s): Remove a standard transmission from a truck chassis using shop equipment, power tools, and hand tools.

ASE Education Foundation Correlation

This job sheet addresses the following ASE Education Foundation task(s):

II. **Drive Train**
   C. **Transmission**
   8. Inspect, adjust, and replace transmission covers, rails, forks, levers, bushings, sleeves, detents, interlocks, springs, and lock bolts/safety wires. (TST, MTST) P-2.
   11. Remove and reinstall transmission. (TST, MTST) P-2.

Tools and Materials: A truck equipped with a standard transmission, wheel chocks, jack stands and **hoists**, and a mobile transmission jack. If a COE truck is used, an A-frame chain and block hoist (chain falls) may be used in place of the transmission jack.

Protective Clothing: Standard shop apparel including coveralls or shop coat, safety glasses, and safety footwear.

PROCEDURE
Ensure that the shop LOTO is observed.

Plan the removal of the transmission from the chassis before beginning the operation. The method you select will depend on the type of chassis, location of the engine mounts (which may be on the transmission), and the amount of auxiliary equipment located around the transmission that may have to be removed. In most cases it is unnecessary to drain the lubricant from the transmission prior to removing it. When the reason for pulling the transmission is to replace a clutch, the transmission oil should not be removed. This procedure is a rough description of the transmission R-and-R on a conventional chassis. IMPORTANT: make sure the shop LOTO policy is observed.

1. Put the tractor into park. Block the wheels. Determine if the truck will have to be raised in order to remove the transmission. Many modern trucks have to be significantly raised with a scissor jack to get the transmission out from under the chassis. Check that all the equipment required can be moved into position.

   Task completed _____

2. Remove the shift lever from the transmission and cap the shift tower to prevent dirt from entering the transmission.

   Task completed _____

3. Separate the U-joint from the transmission tailshaft yoke using a U-joint puller. Remove the PTO drive shaft assembly if fitted.

   Task completed _____

4. Disconnect all the electrical connections from the transmission, including the tailshaft speed sensor wires and reverse switch.

   Task completed _____

5. Bleed down the air pressure in the air tanks. Remove only the air line connections that have to be removed in order to pull the transmission.

   Task completed _____

6. Remove the external clutch linkage assembly, including any brackets and return springs. Note the location of springs on the bell crank.

   Task completed _____

7. Check the location of the engine mounts. If the engine mounts are located on the transmission bell housing, ensure that the weight of the engine is fully supported and secure enough to withstand any movement required during the removal process. In the case of flywheel located engine mounts, leave them in place.

   Task completed _____

8. Place the transmission jack under the transmission and bolt the jack deck brackets in place. Chain the transmission onto the jack deck.

   Task completed _____

9. Remove the transmission bell housing bolts. Insert four 6-inch (150 mm) guide studs located at the 2, 4, 8, and 10 o'clock positions.

   Task completed _____

10. Pull the transmission smoothly and evenly away from the flywheel housing. Ensure that the clutch release yoke does not snag on the release bearing.

    Task completed _____

11. Remove the transmission from under the truck chassis, ensuring that if the truck has to be raised, this can be done safely.

    Task completed _____

CAUTION! Never go under a truck chassis unless it is supported mechanically. Great care should be taken whenever raising a truck using scissor jacks or overhead hoists.

STUDENT SELF-EVALUATION

Check	Level	Competency	Comments
	4	Mastered task	
	3	Competent but need further help	
	2	Needed a lot of help	
	0	Did not understand the task	

INSTRUCTOR EVALUATION

Check	Level	Competency	Comments
	4	Mastered task	
	3	Competent but needs further help	
	2	Requires more training	
	0	Unable to perform task	

ONLINE TASKS

Use a search engine and access the following heavy-duty transmission OEMs. Make a note of their product range. See if you can identify the advantages of a triple countershaft transmission over a twin countershaft.

1. Eaton Fuller Roadranger Division
2. Spicer-Dana
3. Volvo-Mack Trucks

STUDY TIPS

Identify 5 key points in Chapter 15. Try to be as brief as possible.

Key point 1 _____

Key point 2 _____

Key point 3 _____

Key point 4 _____

Key point 5 _____

16 Standard Transmission Servicing

Objectives

After reading this chapter, you should be able to:
- Explain the importance of using the correct lubricant and maintaining the correct oil level in a transmission.
- List the preventive maintenance inspections that should be made periodically on a standard transmission.
- Explain how to replace a transmission rear seal.
- Describe the procedure for troubleshooting standard transmissions.
- Identify the causes of some typical transmission performance problems, such as unusual noises, leaks, vibrations, jumping out of gear, and hard shifting.
- Outline the procedure for overhauling a typical twin-countershaft transmission.
- Disassemble and reassemble a transmission auxiliary drive section.
- Disassemble and reassemble a transmission main case.
- Analyze the procedure for performing failure analysis on transmission components.
- Troubleshoot an air shift system.

PRACTICE QUESTIONS

These questions require you to write out the answers rather than select one of four choices. Keep your answers as brief as possible.

1. Name two mechanical problems that can result from overfilling a standard transmission with oil.

2. What is the most common cause of failure in truck standard transmissions?

3. If it is determined that the rear transmission oil seal is leaking, what should you do to repair the problem?

4. What is the most common cause of gear whine?

5. List four possible causes of transmission noise that will occur only when the transmission is in gear.

6. What should you do to determine whether the transmission itself is the cause of a hard shifting problem?

7. List some causes of transmission bearing failure.

8. List two things that you should do to transmission bolts when reassembling a stripped-down transmission.

9. When a transmission either fails to shift range or does so sluggishly, what could be the cause?

10. List some daily checks that should be performed on a transmission to keep it in the best working condition.

JOB SHEET 16.1

Name _____ Station _____ Date _____

Install a Truck Standard Transmission.

Performance Objective(s): Install a standard transmission into a truck chassis using shop equipment, power tools, and hand tools. This procedure completes the R-and-R procedure begun in the previous job sheet. Clutch installation is covered in Chapter 14.

ASE Education Foundation Correlation

This job sheet addresses the following ASE Education Foundation task(s):

II. Drive Train
C. Transmission
1. Inspect transmission shifter and linkage; inspect transmission mounts, insulators, and mounting bolts. (IMMR, TST, MTST) P-1.
2. Inspect transmission for leakage; determine needed action. (IMMR, TST, MTST) P-1.
3. Replace transmission cover plates, gaskets, seals, and cap bolts; inspect seal surfaces and vents; determine needed action. (IMMR, TST, MTST) P-1.
4. Check transmission fluid level and condition; determine needed action. (IMMR, TST, MTST) P-1.
5. Inspect transmission breather; inspect transmission oil filters, coolers, and related components; determine needed action. (IMMR, TST, MTST) P-2.
6. Inspect speedometer components. (IMMR) P-2.
6. Inspect speedometer components; determine needed action. (TST, MTST) P-2.
7. Inspect and test function of REVERSE light, neutral start, and warning device circuits. (IMMR) P-1.
7. Inspect and test function of REVERSE light, NEUTRAL start, and warning device circuits; determine needed action. (TST, MTST) P-1.
8. Inspect, adjust, and replace transmission covers, rails, forks, levers, bushings, sleeves, detents, interlocks, springs, and lock bolts/safety wires. (TST, MTST) P-2.
10. Inspect, test, repair, and/or replace air shift controls, lines, hoses, valves, regulators, filters, and cylinder assemblies. (TST, MTST) P-2.
11. Remove and reinstall transmission. (TST, MTST) P-2.
12. Inspect input shaft, gear, spacers, bearings, retainers, and slingers; determine needed action. (TST, MTST) P-2.
13. Inspect and adjust power take-off (PTO) assemblies, controls, and shafts; determine needed action. (TST, MTST) P-3.
14. Inspect and test transmission temperature gauge, wiring harnesses, and sensor/sending unit; determine needed action. (TST, MTST) P-2.

Tools and Materials: A truck equipped with a standard transmission, wheel chocks, jack stands and hoists, and a mobile transmission jack. If a COE truck is used, an A-frame chain-and-block hoist may be used in place of the transmission jack.

Protective Clothing: Standard shop apparel including coveralls or shop coat, safety glasses, and safety footwear.

PROCEDURE
Ensure that the shop LOTO is observed.

Plan the installation of the transmission into the chassis before beginning the operation. If you took the transmission out, this procedure should generally be reversed. This procedure is a rough description of the transmission R-and-R on a conventional chassis. It really pays off to ensure that the truck and transmission are level before beginning this procedure: raise the back of the truck and place on stands. This way you will not be fighting gravity as you install the transmission through the clutch pack.

1. First check the clutch friction disc alignment. Insert a pilot shaft fully into the clutch assembly so that the pilot bore enters the pilot bearing. The pilot shaft should withdraw without binding or snagging.

Task completed _____

2. Position the transmission under the truck. Place the transmission into gear so that when the yoke is rotated, the transmission input shaft rotates. Raise the transmission so that the input shaft is level with the clutch release bearing bore.

   Task completed _____

3. Eyeball the angle of the engine flywheel assembly and try to set the angle of the transmission identically.

   Task completed _____

4. Align the transmission and move it gently into the clutch assembly. Locate the transmission input shaft splines in the clutch disc splines using the tailshaft yoke on the transmission to rotate the input shaft.

   Task completed _____

5. Move the clutch fork yoke over the release bearing as the clutch is moved into the clutch assembly. Once the yoke passes over the top of the release bearing, the external release lever can be used to help pull the transmission into the bell housing. Ensure that the transmission input shaft fully enters the flywheel pilot bearing and that the clutch bell housing abuts the flywheel housing before installing bolts.

   Task completed _____

CAUTION! Never pull in a binding transmission by using the bell housing bolts. It will almost certainly result in damage to the clutch disc splines or the pilot bearing.

6. Install all the bell housing bolts and torque to specification. Install the clutch release mechanism and check clutch operation before proceeding.

   Task completed _____

7. Reverse the disassembly procedure and finish the transmission installation, making sure that all air and electrical connections are completed.

   Task completed _____

8. Check the transmission oil level and top it off if necessary.

   Task completed _____

9. Adjust the clutch.

   Task completed _____

10. Road test.

    Task completed _____

STUDENT SELF-EVALUATION

Check	Level	Competency	Comments
	4	Mastered task	
	3	Competent but need further help	
	2	Needed a lot of help	
	0	Did not understand the task	

INSTRUCTOR EVALUATION

Check	Level	Competency	Comments
	4	Mastered task	
	3	Competent but needs further help	
	2	Requires more training	
	0	Unable to perform task	

ONLINE TASKS

Use a search engine and research the cost of replacing a rebuilt transmission versus that of a new transmission. The best way to do this is to go into your shop and record the serial number of an actual transmission: Compare the most expensive quote to the least expensive quote. Input the key words from the "Study Tips" section after you have completed it. You should be able to access more recent information than that in the textbook. See if you can identify any emerging trends.

STUDY TIPS

Identify 5 key points in Chapter 16. Try to be as brief as possible.

Key point 1 _____

Key point 2 _____

Key point 3 _____

Key point 4 _____

Key point 5 _____

17 Torque Converters

Objectives

After reading this chapter, you should be able to:
- Outline the function of the torque converter in a vehicle equipped with an automatic transmission.
- Explain how the torque converter is coupled between the crankshaft and the transmission.
- Identify the three main elements of a torque converter torus and describe their roles.
- Define torque multiplication and explain how it is generated in the torque converter.
- Define both rotary and vortex fluid flows and explain how each affects torque converter operation.
- Describe the overrunning clutch, lockup clutch, and variable pitch stators.
- Outline torque converter service and maintenance checks.
- Remove, disassemble, inspect, and reassemble torque converter components.

PRACTICE QUESTIONS

1. The purpose of the _____ is to transfer crankshaft torque to the torque converter assembly.
   a. lockup clutch
   b. flexplate
   c. stator
   d. turbine

2. Which of the following is the input drive member of a torque converter?
   a. stator
   b. turbine
   c. flexplate
   d. impeller

3. Which of the following must always rotate at crankshaft speed when a torque converter is used to drive an automatic transmission?
   a. impeller
   b. turbine
   c. stator
   d. transmission input shaft

4. Technician A says that torque converter coupling phase means the impeller and turbine are running within 10 percent of the same speed. Technician B says that the only time the turbine and the impeller rotate at identical speeds is when the lockup clutch is engaged. Who is correct?
   a. Technician A only
   b. Technician B only
   c. both A and B
   d. neither A nor B

5. Technician A says that all heavy-duty trucks with automatic transmissions use centrifugal lockup converters. Technician B says that a centrifugal lockup clutch consists of a piston, clutch plate, and back plate. Who is correct?
   a. Technician A only
   b. Technician B only
   c. both A and B
   d. neither A nor B
6. Technician A says that if it is necessary to rebuild a stator assembly, the needle bearing should be removed first. Technician B says that cracked stator vanes can be repair-welded using the TIG welding process. Who is correct?
   a. Technician A only
   b. Technician B only
   c. both A and B
   d. neither A nor B
7. What type of stator tends to be found in heavy-duty truck torque converters?
   a. fixed
   b. rotating
   c. hydraulic
   d. pneumatic
8. When does a stator begin to freewheel?
   a. torque multiplication phase
   b. coupling phase
   c. lockup phase
   d. at impeller stall
9. At what pressure should shop air be regulated to test a sealed-weld torque converter?
   a. 5 psi
   b. 25 psi
   c. 75 psi
   d. 100 psi
10. When is vortex flow at a maximum in a torque converter?
    a. full stall
    b. heavy acceleration
    c. coupling phase
    d. lockup phase

JOB SHEET 17.1

Name _____ Station _____ Date _____

Disassemble a Torque Converter.

Performance Objective(s): Disassemble a bolted housing, heavy-duty torque converter.

ASE Education Foundation Correlation

This job sheet addresses the following ASE Education foundation task(s):

II. Drive Train
   C. Transmission
     11. Remove and reinstall transmission. (TST, MTST) P-2.
     15. Inspect operation of automatic transmission, components, and controls; diagnose automatic transmission system problems; determine needed action. (TST) P-2.
     15. Inspect and test operation of automatic transmission, components, and controls; diagnose automatic transmission system problems; determine needed action. (MTST) P-2.

Tools and Materials: Basic shop power and hand tools.

Protective Clothing: Standard shop apparel including coveralls or shop coat, safety glasses, and safety footwear.

PROCEDURE

IMPORTANT: Ensure the shop LOTO procedure is observed. All of the components are number coded to Figure 17–1, so use this as a guide to the disassembly procedure.

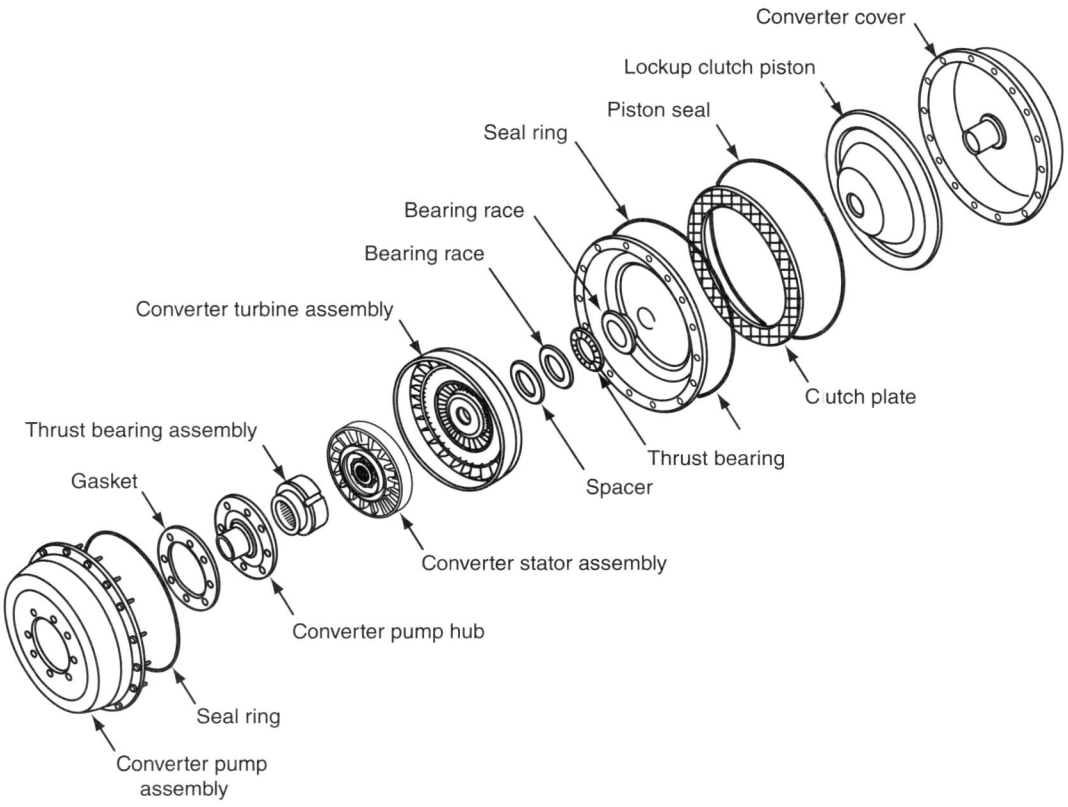

FIGURE 17–1 Components of a typical lockup torque converter.

Torque Converters 209

1. Remove the six rubber id retainers and spacers from the converter cover assembly (6).

    Task completed _____

2. Remove all the nuts (5) from the cover (6).

    Task completed _____

3. Remove the torque converter cover (6) and lockup clutch piston (10) as a unit.

    Task completed _____

4. Place the cover assembly on a worktable with the lockup clutch piston facing up.

    Task completed _____

5. Remove the bearing race (16).

    Task completed _____

6. Compress the center of the piston (10) and remove the snapring (15).

    Task completed _____

7. Turn the cover assembly over (piston down) and knock the cover sharply to remove the piston.

    Task completed _____

8. Remove the seal ring retainer (8) and the seal ring (9) from the cover (6).

    Task completed _____

9. Remove the seal ring (11) from the cover (6).

    Task completed _____

10. Inspect the bushing (7) and remove only if replacement is necessary.

    Task completed _____

11. Remove the lockup clutch plate (12).

    Task completed _____

12. Remove the lockup clutch back plate (14) from the torque converter pump (42).

    Task completed _____

13. Remove the seal ring (13) from the plate (14).

    Task completed _____

14. Remove the thrust bearing assembly (17), bearing race (18), and spacer (19) from the hub of the turbine (20).

    Task completed _____

15. Remove the converter turbine assembly.

    Task completed _____

16. Grasp the stator and the roller race and remove them as a unit.

    Task completed _____

17. Position the stator assembly (24) on the worktable so that the freewheel roller race (32) is upward.

    Task completed _____

18. Remove the roller race by rotating it clockwise while lifting it out of the converter stator.

    Task completed _____

19. Remove the ten rollers (34) and ten springs (33) from the stator assembly (24).

    Task completed _____

20. Wash and flush the needle bearing assembly (35) thoroughly with dry cleaning solvent or mineral spirits.

    Task completed _____

21. Dry the assembly (35), and lubricate it with transmission oil.

    Task completed _____

22. Replace only the freewheel race and rotate the bearing while pressing on the freewheel race.

    a. Is there roughness or binding?

       Yes _____ No _____

       Task completed _____

    b. If not, reuse the needle bearing assembly in the stator and cam assembly.

       Not applicable _____

       Task completed _____

    c. If so, check the needle bearing end of the flywheel race for a smooth finish.

       Not applicable _____

       Task completed _____

    d. Replace the freewheel race if the bearing end is scratched or contains chatter marks.

       Not applicable _____

       Task completed _____

    e. If only the needle bearing assembly requires replacement, remove it carefully to avoid nicking the aluminum bore in which it is held.

       Not applicable _____

       Task completed _____

    f. Place the new needle bearing assembly, thrust race first, into the aluminum bore or the stator.

       Not applicable _____

       Task completed _____

    g. Use a bearing installer and handle to drive the bearing assembly into the stator until the top of the outer shell is 0.025–0.035 inch (0.635–0.889 mm) above the shoulder in the side plate.

       Not applicable _____

       Task completed _____

23. Remove the needle bearing (35), bearing race (48), and roller bearing (47) from the converter pump hub (44).

    Task completed _____

24. Remove the seal ring (41).

    Task completed _____

25. Flatten the corners of the lock strips (38 or 45).

    Task completed _____

26. Remove the eight bolts (37 or 46).

   Task completed _____

27. Remove the four lock strips from the converter pump hub.

   Task completed _____

28. Remove the converter pump hub and gasket (40) from the pump (42).

   Task completed _____

29. Remove the seal ring.

   Task completed _____

STUDENT SELF-EVALUATION

Check	Level	Competency	Comments
	4	Mastered task	
	3	Competent but need further help	
	2	Needed a lot of help	
	0	Did not understand the task	

INSTRUCTOR EVALUATION

Check	Level	Competency	Comments
	4	Mastered task	
	3	Competent but needs further help	
	2	Requires more training	
	0	Unable to perform task	

ONLINE TASKS

Use a search engine to locate a Web page called Banks Power, and then use the hyperlink entitled "Understanding torque converters." Explain the difference between a torque converter and a fluid coupling.

STUDY TIPS

Identify 5 key points in Chapter 17. Try to be as brief as possible.

   Key point 1 _____

   Key point 2 _____

   Key point 3 _____

   Key point 4 _____

   Key point 5 _____

18 Automatic Transmissions

Objectives

After reading this chapter, you should be able to:
- Identify the components of a simple planetary gearset.
- Explain the operating principles of a planetary geartrain.
- Define a compound planetary gearset and explain how its outputs are managed.
- Describe a multiple-disc hydraulic clutch and explain its role in the operation of an automatic transmission.
- Outline torque path powerflow through typical four- and five-speed automatic transmissions.
- Describe the hydraulic circuits and flows used to control automatic transmission operation.
- List the two types of hydraulic retarders used in Allison automatic transmissions and explain their differences.

PRACTICE QUESTIONS

1. Technician A says that the number of planetary pinions in a carrier depends on the torque load that the gearset is expected to carry. Technician B says that the planetary ring gear is capable of exerting the most leverage on the axis of a planetary gearset. Who is correct?
    a. Technician A only
    b. Technician B only
    c. both A and B
    d. neither A nor B

2. Which of the following is an advantage of a simple planetary gearset?
    a. gear force loads are divided equally
    b. compactness
    c. multiple ratios and reverse capability
    d. all of the above

3. When a planetary carrier drives the sun gear and the ring gear is held stationary, which of the following would result?
    a. slow overdrive
    b. maximum reduction
    c. maximum overdrive
    d. reverse

4. Technician A says that multiple-disc clutches act as braking devices in heavy-duty truck automatic transmissions. Technician B says that multiple-disc clutches act as power transfer devices in heavy-duty automatic transmissions. Who is correct?
   a. Technician A only
   b. Technician B only
   c. both A and B
   d. neither A nor B

5. If a planetary carrier is held, the sun gear is the input, and the ring gear is the output, which of the following would result?
   a. reverse reduction
   b. reverse overdrive
   c. forward reduction
   d. forward overdrive

6. Technician A says that a major difference between an Allison four-speed and an Allison five-speed transmission is that the five-speed has an additional planetary gearset. Technician B says that a planetary gearset is required for each gear range in a heavy-duty automatic transmission. Who is correct?
   a. Technician A only
   b. Technician B only
   c. both A and B
   d. neither A nor B

7. When a low planetary gearset is used, it is located _____.
   a. directly behind the torque converter
   b. in front of the first planetary gearset
   c. in the center of the transmission
   d. at the rear of the transmission

8. In an Allison four-speed automatic transmission, governor pressure is directed to which of the following?
   a. 1–2 shift signal valve
   b. 3–4 shift signal valve
   c. modulated lockup valve
   d. all of the above

9. Technician A says that modulator pressure should be lowest at idle and increase with accelerator pedal travel. Technician B says that the trimmer regulator valve is controlled by modulator pressure. Who is correct?
   a. Technician A only
   b. Technician B only
   c. both A and B
   d. neither A nor B

10. Which type of hydraulic retarder uses a two-stage design that applies retarding force directly to the driveline?
    a. input retarder
    b. output retarder
    c. torque retarder
    d. all of the above

JOB SHEET 18.1

Name _____ Station _____ Date _____

Calculate Planetary Gearset Ratios.

Performance Objective(s): Use a hand-rotated planetary gearset and proof out input-output ratios and rotational direction.

ASE Education Foundation Correlation

This job sheet addresses the following ASE Education Foundation task(s):

II. Drive Train
 C. Transmission
  11. Remove and reinstall transmission. (TST, MTST) P-2.
  15. Inspect operation of automatic transmission, components, and controls; diagnose automatic transmission system problems; determine needed action. (TST) P-2.
  15. Inspect and test operation of automatic transmission, components, and controls; diagnose automatic transmission system problems; determine needed action. (MTST) P-2.

Tools and Materials: A bench-mounted planetary gearset, some chalk, a pencil, and paper

Protective Clothing: Standard shop apparel including coveralls or shop coat, safety glasses, and safety footwear

PROCEDURE

Use the following chart and a hand-rotated planetary gearset to proof out input-output ratios and direction of rotation.

Sun Gear	Carrier	Ring Gear	Direction	Ratio
Input	Output	Hold		
Hold	Output	Input		
Output	Input	Hold		
Hold	Input	Output		
Input	Hold	Output		
Output	Hold	Input		
Input	Hold	Hold		

Task completed _____

STUDENT SELF-EVALUATION

Check	Level	Competency	Comments
	4	Mastered task	
	3	Competent but need further help	
	2	Needed a lot of help	
	0	Did not understand the task	

INSTRUCTOR EVALUATION

Check	Level	Competency	Comments
	4	Mastered task	
	3	Competent but needs further help	
	2	Requires more training	
	0	Unable to perform task	

ONLINE TASKS

Get on the "howstuffworks" and Wikipedia Web pages and check out what they have to say about planetary gearsets. Compare this with what you learned in Chapter 18. Make a note of three other vehicle components that make use of a planetary gearset other than an automatic transmission:

1. _____

2. _____

3. _____

STUDY TIPS

Identify 5 key points in Chapter 18. Try to be as brief as possible.

Key point 1 _____

Key point 2 _____

Key point 3 _____

Key point 4 _____

Key point 5 _____

19 Automatic Transmission Maintenance

Objectives

After reading this chapter, you should be able to:

- Perform hot and cold transmission oil level checks.
- Identify the types of hydraulic fluid used in truck automatic transmissions.
- Change automatic transmission oil and filters.
- Inspect transmission oil for signs of contamination.
- Adjust the manual gear selector linkage, mechanical modulator control linkage, and air modulator control on a truck automatic transmission.
- Perform a transmission stall test.
- Perform engine speed and vehicle speed shift point tests.
- Describe basic transmission test stand procedure.
- Test the transmission valve body.
- Summarize some basic inspection and troubleshooting procedures for automatic transmissions.

PRACTICE QUESTIONS

1. Technician A says that automatic transmission oil levels must be checked at both hot and cold oil levels. Technician B says that the hot check should be made when the transmission oil temperature is 140°F (60°C). Who is correct?
   a. Technician A only
   b. Technician B only
   c. both A and B
   d. neither A nor B

2. Oil and filter change intervals for on-highway trucks should be:
   a. 6,000 miles (9,656 kilometers) or 3 months
   b. 12,000 miles (19,312 kilometers) or 6 months
   c. 25,000 miles (40,233 kilometers) or 12 months
   d. lubed for life

3. Technician A says that the governor feed filter should be changed at every oil/filter change. Technician B says that all the transmission bearings should be replaced if metal contaminants are found in the oil pan. Who is correct?
   a. Technician A only
   b. Technician B only
   c. both A and B
   d. neither A nor B

4. Technician A says that an auxiliary filter may be installed in the oil return line between the oil cooler and the transmission. Technician B says that an auxiliary filter is usually installed following a transmission failure. Who is correct?
   a. Technician A only
   b. Technician B only
   c. both A and B
   d. neither A nor B

5. Technician A says that a stall test should not exceed 1 minute. Technician B says that converter-out temperature should not exceed 300°F (150°C). Who is correct?
   a. Technician A only
   b. Technician B only
   c. both A and B
   d. neither A nor B

6. Which of the following determines when an automatic shift takes place?
   a. shift signal valve spring pressure
   b. modulator spring force
   c. main pressure
   d. both a and b

7. Which of the following should be used as the base reference speed for making adjustments on an automatic transmission?
   a. engine rated speed
   b. engine idle speed
   c. engine high idle speed
   d. engine peak torque speed

8. Technician A says that a 1–2 shift on a four-speed automatic should occur at approximately 2 mph (3 kph) below the top speed in first gear. Technician B says that the 2–3 shift on a four-speed automatic should occur at approximately 2 mph (3 kph) above the top speed for second gear. Who is correct?
   a. Technician A only
   b. Technician B only
   c. both A and B
   d. neither A nor B

9. Which of the following would not be a cause of shift cycling?
   a. out-of-adjustment manual selector
   b. improperly adjusted mechanical modulator
   c. sticking valve in the governor
   d. improper shift valve adjustment

10. Which of the following would not be the cause of premature upshifts?
    a. modulator valve actuating rod not installed
    b. improperly adjusted mechanical modulator
    c. sticking valve in the governor
    d. improper shift valve adjustment

JOB SHEET 19.1

Name _____ Station _____ Date _____

Change Oil and Filter on an Allison Standard Oil Pan Transmission.

Performance Objective(s): Become familiar with the procedure for changing the oil and filter on an Allison hydro-mechanical transmission with a standard oil pan.

ASE Education Foundation Correlation

This job sheet addresses the following ASE Education Foundation task(s):

II. **Drive Train**
   C. **Transmission**
   2. Inspect transmission for leakage; determine needed action. (IMMR, TST, MTST) P-1.
   3. Replace transmission cover plates, gaskets, seals, and cap bolts; inspect seal surfaces and vents; determine needed action. (IMMR, TST, MTST) P-1.
   4. Check transmission fluid level and condition; determine needed action. (IMMR, TST, MTST) P-1.
   5. Inspect transmission breather; inspect transmission oil filters, coolers and related components; determine needed action. (IMMR, TST, MTST) P-2.
   12. Inspect input shaft, gear, spacers, bearings, retainers, and slingers; determine needed action. (TST, MTST) P-2.
   15. Inspect operation of automatic transmission, components, and controls; diagnose automatic transmission system problems; determine needed action. (TST) P-2.
   15. Inspect and test operation of automatic transmission, components, and controls; diagnose automatic transmission system problems; determine needed action. (MTST) P-2.

Tools and Materials: Shop hand tools, power tools, Allison gaskets and filter, drain pan, and torque wrenches

Protective Clothing: Standard shop apparel including coveralls or shop coat, safety glasses, and safety footwear

PROCEDURE
Ensure that the shop LOTO is observed.

1. Position the drain pan under the transmission oil pan.

   Task completed _____

2. Remove the oil drain plug and gasket from the right side of the oil pan.

   Task completed _____

3. Allow the oil to drain.

   Task completed _____

4. Remove the oil pan, gasket, and oil filter tube from the transmission.

   Task completed _____

5. Discard the gasket.

   Task completed _____

6. Clean the oil pan.

   Task completed _____

7. Remove the screw that retains the oil filter.

   Task completed _____

8. Remove the filter and discard it.

   Task completed _____

9. Install the filter tube into the new filter assembly.

   Task completed _____

10. Install a new seal ring onto the filter tube.

    Task completed _____

11. Lubricate the seal ring with transmission oil.

    Task completed _____

12. Install the new oil filter, inserting the filter tube into the hole at the bottom of the transmission.

    Task completed _____

13. Secure the filter with the screw tightened 10–15 lb-ft. (14–20 N·m) of torque.

    Task completed _____

14. Place the oil pan gasket onto the oil pan.

    Task completed _____

15. Install the oil pan and gasket, carefully guiding them into place. Be certain that dirt or other material does not enter the pan.

    Task completed _____

16. Turn in the oil pan retaining screws by hand and then torque all retaining screws to specifications.

    Task completed _____

17. Install the filter tube at the side of the pan and tighten the tube fitting to spec.

    Task completed _____

18. Install the drain plug and gasket and tighten the plug.

    Task completed _____

19. Fill the transmission with oil to the proper level. Follow the manufacturer's specs for oil capacity.

    Task completed _____

20. Recheck the oil level.

    Task completed _____

STUDENT SELF-EVALUATION

Check	Level	Competency	Comments
	4	Mastered task	
	3	Competent but need further help	
	2	Needed a lot of help	
	0	Did not understand the task	

INSTRUCTOR EVALUATION

Check	Level	Competency	Comments
	4	Mastered task	
	3	Competent but needs further help	
	2	Requires more training	
	0	Unable to perform task	

JOB SHEET 19.2

Name _____ Station _____ Date _____

Perform a Stall Test.

Performance Objective(s): Perform a standard stall test on an Allison transmission.

ASE Education Foundation Correlation

This job sheet addresses the following ASE Education Foundation task(s):

II. Drive Train
   C. Transmission
   1. Inspect transmission shifter and linkage; inspect transmission mounts, insulators, and mounting bolts. (IMMR, TST, MTST) P-1.
   2. Inspect transmission for leakage; determine needed action. (IMMR, TST, MTST) P-1.
   4. Check transmission fluid level and condition; determine needed action. (IMMR, TST, MTST) P-1.
   9. Identify causes of transmission noise, shifting concerns, lockup, jumping out-of-gear, overheating, and vibration problems; determine needed repairs. (TST, MTST) P-1.
   14. Inspect and test transmission temperature gauge, wiring harnesses, and sensor/sending unit; determine needed action. (TST, MTST) P-3.
   15. Inspect operation of automatic transmission, components, and controls; diagnose automatic transmission system problems; determine needed action. (TST) P-2.
   15. Inspect and test operation of automatic transmission, components, and controls; diagnose automatic transmission system problems; determine needed action. (MTST) P-2.

Tools and Materials: None

Protective Clothing: Standard shop apparel including coveralls or shop coat, safety glasses, and safety footwear

CAUTION! Before attempting to perform a stall test, check the OEM service literature for any specific requirements for the system you are testing. An improperly executed stall test can damage the engine or transmission, or both.

PROCEDURE
Ensure that the shop LOTO is observed.

1. Connect a tachometer of known accuracy to the engine.

   Task completed _____

2. Install a temperature probe into the converter-out (to cooler) line.

   Task completed _____

3. Start the engine and bring the transmission oil up to its normal operating range temperature (160–200°F/70–93°C).

   Task completed _____

4. Firmly apply the parking and service brakes.

   Task completed _____

5. Block the drive wheels to prevent movement during the test.

   Task completed _____

6. Chain the vehicle to the floor with floor pegs.

   Task completed _____

7. Check that all personnel are out of the vehicle travel path in the event of brake or restraint failure.

   Task completed _____

8. Shift the selector control to any forward range with the exception of D1 or DR (deep ratio) models.

   Task completed _____

9. Accelerate the engine to wide-open throttle.

Task completed _____

10. When the converter-out temperature reaches 255°F (124°C) minimum, record the engine rpm while at wide-open throttle.

Task completed _____

11. Release the throttle and shift to neutral.

Task completed _____

12. Note the converter-out temperature.

Task completed _____

13. Run the engine at 1,200–1,500 rpm for 2 minutes while the transmission remains in neutral.

Task completed _____

14. Record the converter-out temperature at the end of this period.

Task completed _____

15. Evaluate the results.

Task completed _____

STUDENT SELF-EVALUATION

Check	Level	Competency	Comments
	4	Mastered task	
	3	Competent but need further help	
	2	Needed a lot of help	
	0	Did not understand the task	

INSTRUCTOR EVALUATION

Check	Level	Competency	Comments
	4	Mastered task	
	3	Competent but needs further help	
	2	Requires more training	
	0	Unable to perform task	

ONLINE TASKS

Use a search engine and get on the "Banks Power" Web page and then search for the following hyperlink: "Understanding stall tests." Make a note of any differences between performing this test on a truck versus that used to test an automobile.

STUDY TIPS

Identify 5 key points in Chapter 19. Try to be as brief as possible.

Key point 1 _____

Key point 2 _____

Key point 3 _____

Key point 4 _____

Key point 5 _____

20 Automated Manual and Hybrid Transmissions

Objectives

After reading this chapter, you should be able to:

- Explain how a standard mechanical transmission is adapted for automated shifting in three-pedal and two-pedal systems.
- Identify different OEM automated transmissions and interpret some of the serial number codes used.
- Describe the hardware changes that differentiate a standard Roadranger twin-countershaft transmission from its electronically automated version.
- Outline the electronic circuit components that are used to manage AutoShift and UltraShift transmissions.
- Outline the main box and auxiliary section actuator components required for AutoShift and UltraShift electronically automated manual transmissions.
- Describe how the electronic circuit components work together to perform the system functions and outline the role played by the transmission ECU.
- Explain exactly how a shift takes place in an automated manual transmission.
- Perform some basic diagnostic troubleshooting on automated manual transmission electronics.
- Identify the DT-12, Mercedes-Benz AGS, ZF Meritor SureShift, and FreedomLine transmissions.
- Identify hybrid-electric and hybrid-hydraulic transmissions.

PRACTICE QUESTIONS

1. In the prefix of an Eaton Fuller AutoShift transmission code, what does the letter A represent?
    a. automatic
    b. autoshift
    c. automated
    d. active

2. Which data bus would be required to interface a FreedomLine transmission with the engine electronics?
    a. J1930
    b. J2006
    c. J1667
    d. J1939

3. What electrical operating principle does the shift lever use to signal a range request to the transmission ECU?
    a. ground-out
    b. potentiometer
    c. variable capacitance
    d. hall effect

4. In a two-pedal automated transmission system, which of the following pedals is eliminated?
    a. accelerator
    b. service application
    c. clutch
    d. parking brake

5. Which of the following inputs is required by electronically automated standard transmission electronics to manage shifting logic?
   a. road speed
   b. throttle position status
   c. transmission input shaft rpm
   d. all of the above

6. What technology permits AMT electronics to communicate with the engine electronics?
   a. proprietary bus line
   b. multiplexing
   c. optical harness
   d. electromechanical relay

7. When an AMT defaults to one-speed fallback, how is transmission performance affected?
   a. Shifts are inhibited to one range up, one range down.
   b. Road speed defaults to the speed at which the failure mode was logged.
   c. Road speed locks to the set speed.
   d. All shifts are inhibited, and it remains locked in one gear ratio.

8. If an AutoShift fault code 71 was logged, which of the following has likely occurred?
   a. stuck engaged in one gear
   b. failed to engage gear
   c. failed to synchronize engagement
   d. stuck inertia brake solenoid

9. In a three-pedal, automated transmission, when should the driver use the clutch?
   a. for starting only
   b. for stopping and starting
   c. only for downshifts
   d. only for skip-shifts

10. Which of the following makes a two-pedal, AMT *different* from a three-pedal, automated transmission system?
    a. There is no ECU on a two-pedal system.
    b. The three-pedal system is more advanced.
    c. The two-pedal system has no clutch pedal.
    d. There are no countershafts on a two-pedal system.

JOB SHEET 20.1

Name _____ Station _____ Date _____

Retrieve Logged Codes and Erase from AutoShift.

Performance Objective(s): Use an EST loaded with Eaton software to retrieve a logged inactive code and erase.

ASE Education Foundation Correlation

This job sheet addresses the following ASE Education Foundation task(s):

II. **Drive Train**
   A. **General**
   1. Research vehicle service information, including fluid type, vehicle service history, service precautions, and technical service bulletins. (IMMR, TST, MTST) P-1.

   C. **Transmission**
   16. Inspect operation of automated mechanical transmission, components, and controls; diagnose automated mechanical transmission system problems; determine needed action. (TST) P-2.
   16. Inspect and test operation of automated mechanical transmission, components, and controls; diagnose automated mechanical transmission system problems; determine needed action. (MTST) P-2.

Tools and Materials: A truck equipped with an AMT transmission, an EST loaded with the appropriate OEM software

Protective Clothing: Standard shop apparel including coveralls or shop coat, safety glasses, and safety footwear

PROCEDURE
Ensure that the shop LOTO is observed.

1. Start the engine and briefly open the TPS circuit, making sure that it is properly connected again. When the TPS is disconnected, the engine may default to other than an idle rpm depending on the engine in the chassis and how its failure strategy has been programmed.

   Task completed _____

2. Switch off the engine and connect an EST to the chassis data connector.

   Task completed _____

3. Select the transmission SA/MID.

   Task completed _____

4. Scroll to read the active and inactive codes. There should be no active codes and only the TPS inactive code on display.

   Task completed _____

5. Scroll through the procedure to erase inactive codes.

   Task completed _____

6. Next select the engine SA/MID. Erase the logged TPS code from the engine inactive code log.

   Task completed _____

7. Start the engine and ensure that the truck functions properly.

   Task completed _____

STUDENT SELF-EVALUATION

Check	Level	Competency	Comments
	4	Mastered task	
	3	Competent but need further help	
	2	Needed a lot of help	
	0	Did not understand the task	

INSTRUCTOR EVALUATION

Check	Level	Competency	Comments
	4	Mastered task	
	3	Competent but needs further help	
	2	Requires more training	
	0	Unable to perform task	

ONLINE TASKS

Use a search engine and research automated transmissions using the following key words:

1. UltraShift
2. AutoShift
3. FreedomLine
4. Mercedes-Benz AGS

List the advantages of two-pedal automated transmissions versus three-pedal systems.

STUDY TIPS

Identify 5 key points in Chapter 20. Try to be as brief as possible.

Key point 1 _____

Key point 2 _____

Key point 3 _____

Key point 4 _____

Key point 5 _____

21 Electronically Controlled Automatic Transmissions (ECATs)

Objectives

After reading this chapter, you should be able to:

- Identify the Allison and Caterpillar families of electronically controlled, automatic transmissions (ECATs).
- Describe the modular designs used by Allison WT and TC10 transmissions.
- Outline how Allison WT, Allison TC10, and Caterpillar CX transmissions use full authority management electronics to manage shifting and communicate with other vehicle electronic systems.
- Identify some of the service and repair advantages of the modular construction of Allison WT and TC10 transmissions.
- Identify the components used in a Caterpillar CX and note the differences between six- and eight-speed models.
- Describe how the electronic control modules on both Allison and Caterpillar ECATs master the operation of the transmission.
- Define the terms *pulse width modulation, primary modulation,* and *secondary modulation*.
- Describe how base versions of Allison WT and Caterpillar ECATs use interconnected planetary gearsets to stage gearing to provide six forward ranges, reverse, and neutral.
- Outline the differences between Caterpillar CX six- and eight-speed transmissions.
- Describe the integral driveline retarder components and operating principles used in electronic automatic transmissions.
- Outline the function of the dropbox option in one Allison WT model.
- Describe the role of the electrohydraulic controls used in Allison and Caterpillar ECATs.
- Identify the electronic components used in ECATs and classify them as input circuit, processing, and output circuit components.
- Describe how ECAT electronics interface with the J1939 data bus to optimize vehicle performance.
- Describe how the electrohydraulic clutches are controlled.
- Outline the torque paths through WT and TC10 transmissions in each range selected.
- Identify the modules and components of an Allison TC10.
- Describe how diagnostic codes are logged within ECATs and the manner in which they are displayed.
- Outline some maintenance practices used on ECATs.
- Perform some basic diagnostic troubleshooting on Allison and Caterpillar ECATs.
- Calibrate a Caterpillar CX transmission.
- Outline the operation of a Voith transmission.

PRACTICE QUESTIONS

1. Which of the following terms best describes the management system used on Allison CEC transmissions?
   a. hydromechanical
   b. hydromechanical assist
   c. partial authority electronic
   d. full authority electronic

227

2. Which of the following is the intended application of a TC10 ECAT?
   a. city bus
   b. fire truck
   c. garbage packer
   d. linehaul truck

3. In which TC10 gear ratios does converter input torque bypass the main gearbox module?
   a. 1st and low reverse
   b. 2nd and 5th
   c. 4th and 9th
   d. 5th and 10th

4. Technician A says that input torque to the main gearbox module in a TC10 is divided between its twin countershafts. Technician B says that the main planetary in a TC10 is located in the main gearbox. Who is correct?
   a. Technician A only
   b. Technician B only
   c. both A and B
   d. neither A nor B

5. What is Allison diagnostic software known as?
   a. ET
   b. MD
   c. DOC
   d. WTEC

6. Technician A says that Allison input shaft rpm data can be viewed in the DOC StripChart field. Technician B says that Allison output shaft rpm data can be viewed in the DOC StripChart field. Who is correct?
   a. Technician A only
   b. Technician B only
   c. both A and B
   d. neither A nor B

7. What software is required to troubleshoot a CX31 transmission?
   a. DOC
   b. ET
   c. E-Tech II
   d. MD

8. Technician A says using AT-1 synthetic fluid in a CX-31 transmission considerably shortens service life. Technician B says that using TranSynd synthetic fluid in Allison transmissions considerably extends service life. Who is correct?
   a. Technician A only
   b. Technician B only
   c. both A and B
   d. neither A nor B

9. What is the function of the torsional damper on a WT torque converter?
   a. dampens engine thrust forces
   b. dampens drive shaft torsionals
   c. smoothes engine torsionals
   d. smoothes shift torque

10. What is the term used by Allison to describe the subsystems that make up the WT transmission?
    a. subcircuits
    b. modules
    c. subcomponents
    d. compounds

11. How many gearbox modules are there in an Allison WT transmission?
    a. one
    b. three
    c. five
    d. six

12. Which of the following is classified by Allison as an input module?
    a. torque converter module
    b. rotating clutch module
    c. main shaft module
    d. retarder module

13. Which of the following correctly describes the means used to control the retarder solenoids?
    a. normally open, pulse width modulated
    b. normally closed, pulse width modulated
    c. normally open, analog
    d. normally closed, analog

14. Technician A says that a WT VIM has at least two relays, one of which signals the reverse warning circuit. Technician B says that a WT VIM may contain up to six relays. Who is correct?
    a. Technician A only
    b. Technician B only
    c. both A and B
    d. neither A nor B

15. Technician A says that the WT tailshaft sensor uses a 100-tooth reluctor wheel. Technician B says that the J1939 data bus allows WT electronics to share the TPS signal with the engine. Who is correct?
    a. Technician A only
    b. Technician B only
    c. both A and B
    d. neither A nor B

16. Technician A says that in a PWM signal, the percentage of on-off time in one cycle determines how switching outcomes are managed. Technician B says that the cycle frequency will determine how duty cycle is managed. Who is correct?
    a. Technician A only
    b. Technician B only
    c. both A and B
    d. neither A nor B

17. Which solenoids have to be energized when a WT transmission is in fourth gear?
    a. A, B, E
    b. B, E, G
    c. A, D, F
    d. F, G

JOB SHEET 21.1

Name _____ Station _____ Date _____

Perform a Snapshot Test on an Allison WT Transmission.

Performance Objective(s): Use an electronic service tool (EST) loaded with the appropriate Allison software to perform a snapshot test during a road test.

ASE Education Foundation Correlation

This job sheet addresses the following ASE Education Foundation task(s):

II. Drive Train
  A. General
   1. Research vehicle service information, including fluid type, vehicle service history, service precautions, and technical service bulletins. (IMMR, TST, MTST) P-1.
  C. Transmission
   15. Inspect operation of automatic transmission, components, and controls; diagnose automatic transmission system problems; determine needed action. (TST) P-2.
   15. Inspect and test operation of automatic transmission, components, and controls; diagnose automatic transmission system problems; determine needed action. (MTST) P-2.

Tools and Materials: Truck equipped with a CEC transmission, EST loaded with Allison software, shop hand and power tools

Protective Clothing: Standard shop apparel including coveralls or shop coat, safety glasses, and safety footwear

PROCEDURE
Ensure that the shop LOTO is observed.

If the "road test" is to be performed on a highway, ensure that you are properly licensed and that you fully understand how to operate the truck.

1. Perform a circle-check and ensure that the vehicle is in safe condition for a road test or even to be moved within a compound.

   Task completed _____

2. Connect the EST to the chassis data connector and scroll the menu to select the transmission MID.

   Task completed _____

3. Get into the snapshot test menu. Select the parameters to be monitored and the time intervals.

   Task completed _____

4. Select a manual trigger.

   Task completed _____

5. Perform the test sequence, and, at some point, key the manual trigger. It is not necessary to drive any distance to perform this test.

   Task completed _____

6. Park the vehicle. Print the snapshot data. Discuss how snapshot tests can be useful in identifying intermittent faults.

   Task completed _____

STUDENT SELF-EVALUATION

Check	Level	Competency	Comments
	4	Mastered task	
	3	Competent but need further help	
	2	Needed a lot of help	
	0	Did not understand the task	

INSTRUCTOR EVALUATION

Check	Level	Competency	Comments
	4	Mastered task	
	3	Competent but needs further help	
	2	Requires more training	
	0	Unable to perform task	

JOB SHEET 21.2

Name _____ Station _____ Date _____

Test an Allison WT Electrohydraulic Valve Body.

Performance Objective(s): Observe the testing procedure used to verify the performance of an Allison WT electrohydraulic valve body.

ASE Education Foundation Correlation

This job sheet addresses the following ASE Education Foundation task(s):

II. Drive Train
   A. General
   1. Research vehicle service information, including fluid type, vehicle service history, service precautions, and technical service bulletins. (IMMR, TST, MTST) P-1.
   C. Transmission
   15. Inspect operation of automatic transmission, components, and controls; diagnose automatic transmission system problems; determine needed action. (TST) P-2.
   15. Inspect and test operation of automatic transmission, components, and controls; diagnose automatic transmission system problems; determine needed action. (MTST) P-2.

Tools and Materials: An Allison transmission dealership or a transit corporation shop that performs transmission work in-house

Protective Clothing: Standard shop apparel including coveralls or shop coat, safety glasses, and safety footwear

PROCEDURE
Ensure that the shop LOTO is observed.

The testing of an electrohydraulic valve body must be done in a facility with an Allison test fixture. This means locating an Allison transmission dealership or transit bus operation with the appropriate equipment, because few training institutions will be in possession of this equipment. The objective of the exercise is to correlate the test shift positions with the hydraulic schematics and powerflow diagrams in the textbook.

1. Locate a facility with the capability to test Allison electrohydraulic valve bodies and make arrangements to have it demonstrated.

   Task completed _____

2. Make photocopies of the WT hydraulic and powerflow schematics in the textbook. Note on each photocopy the clutch apply, solenoid actuation, and latching conditions in the transmission.

   Task completed _____

3. Attend the demonstration with the photocopies and correlate the test profile with the data you have compiled.

   Task completed _____

STUDENT SELF-EVALUATION

Check	Level	Competency	Comments
	4	Mastered task	
	3	Competent but need further help	
	2	Needed a lot of help	
	0	Did not understand the task	

INSTRUCTOR EVALUATION

Check	Level	Competency	Comments
	4	Mastered task	
	3	Competent but needs further help	
	2	Requires more training	
	0	Unable to perform task	

JOB SHEET 21.3

Name _____ Station _____ Date _____

Calibrate a Caterpillar CX31 Transmission.

Performance Objective(s): Perform the calibration procedure required on a CX31 transmission when any major component is rebuilt or replaced.

ASE Education Foundation Correlation

This job sheet addresses the following ASE Education Foundation task(s):

II. Drive Train
  A. General
  1. Research vehicle service information, including fluid type, vehicle service history, service precautions, and technical service bulletins. (IMMR, TST, MTST) P-1.
  C. Transmission
  15. Inspect operation of automatic transmission, components, and controls; diagnose automatic transmission system problems; determine needed action. (TST) P-2.
  15. Inspect and test operation of automatic transmission, components, and controls; diagnose automatic transmission system problems; determine needed action. (MTST) P-2.

Tools and Materials: A Caterpillar CX transmission in a truck chassis and Caterpillar ET

Protective Clothing: Standard shop apparel including coveralls or shop coat, safety glasses, and safety footwear

CALIBRATION PROCEDURE

To calibrate a CX transmission, the vehicle must be driven through an extensive road test while the technician/driver responds to prompts from the electronic technician (ET). The following steps map out the sequence required for CX28 and CX31 but they should not be attempted without the appropriate Caterpillar training and software:

1. Check and correct the transmission oil level.

   Task completed _____

2. Connect an EST to J1939 and launch ET.

   Task completed _____

3. Select <configuration> from the menu options, then <vehicle configuration parameters> followed by <transmission calibration configuration>. Program the calibration to <all clutches>.

   Task completed _____

4. Cycle power to the ECU in this sequence: key-on, key-off, key-on.

   Task completed _____

5. Drive the unit until the oil temperature is at normal operating temperature. This means above 60°C (140°F) and below 100°C (212°F).

   Task completed _____

6. In a lot away from a public highway, release the parking brakes and apply the service brakes, and then shift from N to D then through N to R. Repeat this sequence at least five times or until each shift change feels similar.

   Task completed _____

7. In a safe place, accelerate the vehicle so that it shifts to second gear while maintaining an engine speed of 1,200 rpm. This will prompt a series of shifts from second to first and back in 4-second intervals; the accelerator pedal should be held to run the engine at 1,200 rpm. Repeat this process at least three times or until the shifts feel smooth.

Task completed _____

8. Drive the vehicle at around 50 percent throttle to generate upshifts from first to fifth gears. Then coast the vehicle, permitting the transmission to downshift back to first gear. Moderate braking may be used during the downshift sequence.

Task completed _____

9. Accelerate through upshifts to fifth gear again, and then accelerate sufficiently to prompt a downshift back to fourth gear. After allowing a shift back to fifth gear again, release the accelerator pedal and allow the vehicle to coast back to fourth gear. The range for the shift point is 1,200–1,400 rpm on an engine programmed to achieve peak torque at 1,200 rpm. Depending on the engine peak torque rpm, the shift point will rise proportionally with peak torque rpm. This shift sequence should be repeated for a minimum of five times or until there is no noticeable change in shift quality.

Task completed _____

10. Road test the vehicle ensuring that the transmission upshifts and downshifts through all six speeds smoothly. If shifting continues to be harsh, troubleshoot the transmission using ET.

Task completed _____

STUDENT SELF-EVALUATION

Check	Level	Competency	Comments
	4	Mastered task	
	3	Competent but need further help	
	2	Needed a lot of help	
	0	Did not understand the task	

INSTRUCTOR EVALUATION

Check	Level	Competency	Comments
	4	Mastered task	
	3	Competent but needs further help	
	2	Requires more training	
	0	Unable to perform task	

ONLINE TASKS

Use a search engine and research which electronic automatic transmissions (Allison by family 1000/2000, 3000/4000, or TC10; Caterpillar by family CX28, CX31, and CX35) can be spec'd to the engines in the list that follows. You may find you have a couple of choices for each, but the BHP and torque specs should match.

1. Caterpillar C13
2. Navistar MaxxForce 11
3. Volvo VDE13
4. Detroit Diesel DD15

STUDY TIPS

Identify 5 key points in Chapter 21. Try to be as brief as possible.

Key point 1 _____

Key point 2 _____

Key point 3 _____

Key point 4 _____

Key point 5 _____

22 Driveshaft Assemblies

Objectives

After reading this chapter, you should be able to:
- Identify the components in a truck driveline.
- Explain the procedures for inspecting, lubricating, and replacing a universal joint.
- Describe the various types of universal joint wear.
- Outline the procedure for sourcing chassis driveline vibration.
- Remove and replace a driveshaft universal joint.
- Define and explain the importance of driveshaft phasing.
- Explain the importance of driveline working angles and how to calculate them.
- Phase a heavy-duty driveshaft assembly.
- Describe the procedure for balancing a driveshaft.

PRACTICE QUESTIONS

1. Technician A says that oversized driveshafts can place excessive strain on transmission and drive axle components. Technician B says that driveshaft slip splines are often coated with phosphate or polymer skins. Who is correct?
   a. Technician A only
   b. Technician B only
   c. both A and B
   d. neither A nor B

2. When lubricating U-joints, which of the following is correct practice?
   a. Pump grease in until it exits from one trunnion seal.
   b. Disassemble the U-joint and pack the needle bearings with grease.
   c. Use a torch to warm the U-joint prior to lubricating it at winter temperatures.
   d. Pump grease until it exits from all four trunnion seals.

3. Technician A says that the yokes in a driveshaft propeller must be aligned when assembled at the slip splines. Technician B says that exceeding maximum U-joint working angles can greatly shorten service life. Who is correct?
   a. Technician A only
   b. Technician B only
   c. both A and B
   d. neither A nor B

4. Technician A says that driveshaft length is the main factor in determining the U-joint working angles. Technician B says that driveshaft speed fluctuations can be smoothed by setting the U-joints 90 degrees out-of-phase. Who is correct?
   a. Technician A only
   b. Technician B only
   c. both A and B
   d. neither A nor B

5. Which is the recommended type of grease to be used in heavy-duty truck U-joints?
   a. EP grease meeting NLGI grade 1 specification
   b. EP grease meeting NLGI grade 2 specification
   c. GL lube meeting API GL 5 specification
   d. A and B

6. When greasing a slip joint, which of the following is the correct procedure?
   a. Pump in grease until it exits the spline seal.
   b. Pump in grease until it exits the pressure relief port.
   c. Close off the pressure relief port and pump in grease until it just begins to exit the spline seals.
   d. Disassemble the propeller shaft and pack the slip splines with grease.

7. Which of the following would be a likely cause of galling on trunnion races?
   a. friction caused by improper lubrication
   b. moisture corrosion
   c. out-of-phase operation
   d. vehicle operated in lug

8. What is the general specification for maximum transmission yoke runout?
   a. 0.001 inch (0.025 mm)
   b. 0.005 inch (0.127 mm)
   c. 0.020 inch (0.508 mm)
   d. 0.050 inch (1.27 mm)

9. What is the maximum ovality specification for a heavy-duty driveshaft?
   a. 0.001 inch (0.025 mm)
   b. 0.010 inch (0.25 mm)
   c. 0.025 inch (0.635 mm)
   d. 0.050 inch (1.27 mm)

10. Which of the following could cause the U-joint working angles to change?
    a. lengthening or shortening a chassis and driveshaft
    b. worn-out engine mounts
    c. worn-out suspension bushings
    d. all of the above

JOB SHEET 22.1

Name _____ Station _____ Date _____

Replace a Universal Joint (U-Joint).

Performance Objective(s): Become familiar with the procedure required to replace a standard U-joint.

ASE Education Foundation Correlation

This job sheet addresses the following ASE Education Foundation task(s):

II. Drive Train
   D. Driveshaft and Universal Joints
      1. Inspect, service, and/or replace driveshafts, slip joints, yokes, drive flanges, support bearings, universal joints, boots, seals, and retaining/mounting hardware; check phasing of all shafts. (IMMR, TST, MTST) P-1.
      3. Inspect driveshaft center support bearings and mounts; determine needed action. (TST, MTST) P-1.

Tools and Materials: Standard shop hand tools, power tools, a U-joint puller, a bench vise, a U-joint kit, and antiseize compound

Protective Clothing: Standard shop apparel including coveralls or shop coat, safety glasses, and safety footwear

PROCEDURE
Ensure that the shop LOTO is observed.

1. Use a light hammer and chisel to bend the tabs of the lockstrap away from the bolt heads.

    Task completed _____

2. Remove the four bolts connecting the two bearing cups to the yokes.

    Task completed _____

3. Clamp the U-joint puller to each bearing assembly and pull each bearing cup from the yoke.

    Task completed _____

4. Free the trunnion from the end yoke by tilting the U-joint until the cross clears the yoke bore.

    Task completed _____

5. Collapse the driveshaft and lower the end to the ground.

    Task completed _____

6. Remove the U-joint on the opposite end of the driveshaft.

    Task completed _____

7. Place each end of the driveshaft, less cross and bearing kits, in a bench vise.

    Task completed _____

8. Remove the new cross and bearings from the packaging and separate all four cups from the cross.

    Task completed _____

9. Rotate the cross to inspect for the presence of the one-way check valve in each lube hole of all four trunnions.

    Task completed _____

10. Position the cross into the end yoke with its lube zerk fitting in line as near as possible with the slip spline; lube zerk fitting.

    Task completed _____

11. Apply antiseize compound to the outside of each bearing cup.

    Task completed _____

12. Move one end of the cross to cause a trunnion to project through the yoke eye, and then do the same on the other side.

    Task completed _____

13. Place a bearing cup over one trunnion and align it to the yoke eye.

    Task completed _____

14. While holding the trunnion in alignment with the yoke eye, press the bearing assembly flush to the face of the end yoke by hand.

    Task completed _____

15. If the bearing assembly binds in the cross hole, lightly tap with a nylon hammer directly in the center of the bearing assembly plate. Do not tap the outer edges of the bearing cup cap.

    Task completed _____

16. When the bearing assembly is seated, put the lock plate in place.

    Task completed _____

17. Insert the capscrews that are provided with the kit through the capscrew holes in both the lockstrap and bearing cup cap assembly.

    Task completed _____

18. Thread the capscrews by hand or with a wrench into the bolt holes in the yoke, but do not torque down the bolts.

    Task completed _____

19. Move the cross laterally to the opposite side and through the cross beyond the machined surface of the yoke face.

    Task completed _____

20. Place a bearing cup over the cross trunnion and slide it into the cross hole, seating the cap plate to the face of the yoke.

    Task completed _____

21. Put the lock plate in place.

    Task completed _____

22. Thread the bolts by hand or wrench into the tapped holes in the yoke.

    Task completed _____

23. Repeat the process at the opposite end of the driveshaft.

    Task completed _____

24. Torque all capscrews and lock them by bending the lockstrap tabs into the capscrew flats.

    Task completed _____

STUDENT SELF-EVALUATION

Check	Level	Competency	Comments
	4	Mastered task	
	3	Competent but need further help	
	2	Needed a lot of help	
	0	Did not understand the task	

INSTRUCTOR EVALUATION

Check	Level	Competency	Comments
	4	Mastered task	
	3	Competent but needs further help	
	2	Requires more training	
	0	Unable to perform task	

ONLINE TASKS

Log on to http://www.jakebrake.com and check out the FAQ link on the site. List the advantages and disadvantages of using a driveline retarder versus an internal engine brake.

Advantages:

1. _____
2. _____
3. _____

Disadvantages:

1. _____
2. _____
3. _____

STUDY TIPS

Identify 5 key points in Chapter 22. Try to be as brief as possible.

Key point 1 _____

Key point 2 _____

Key point 3 _____

Key point 4 _____

Key point 5 _____

23 Heavy-Duty Truck Axles

Objectives

After reading this chapter, you should be able to:

- Identify the types of axles used on trucks and trailers.
- Define the terms dead axle, live axle, pusher axle, and tag axle.
- Outline the construction of a drive axle carrier assembly.
- Explain how a pinion and crown gear set changes the direction of powerflow.
- Describe differential action and list the reasons it is required.
- Identify the components required to create differential action.
- Describe the operation of the various drive axle configurations.
- Identify the components used in an interaxle differential or power divider.
- Explain how an interaxle differential lock functions.
- Define the term spinout and explain how it is caused.
- Trace the powerflow path through different types of differential carriers.

PRACTICE QUESTIONS

1. Technician A says that an interaxle differential lock should be used to lock out differential action between the wheels on either side of a drive axle. Technician B says that an activated interaxle differential locks out differential action between the forward and rear axles on a tandem drive configuration. Who is correct?
   a. Technician A only
   b. Technician B only
   c. both A and B
   d. neither A nor B

2. Which of the following uses two gear sets to provide greater gear reduction and higher torque at the differential carrier?
   a. double reduction axle
   b. two-speed axle
   c. tandem drive axle
   d. differential gear set

3. Technician A says that kerosene should be used to clean and flush a differential carrier housing. Technician B says that it is always recommended that the factory fill axle lubricant be drained and refilled after the first 3,000 linehaul miles (5,000 km). Who is correct?
   a. Technician A only
   b. Technician B only
   c. both A and B
   d. neither A nor B

4. Technician A says that in a tandem drive axle with an interaxle differential, if three wheels are held and one is allowed to spin out, the rpm of the spinning wheel would be four times greater than if all four wheels were turning. Technician B says that wheel spinout is a major cause of differential carrier failure. Who is correct?
   a. Technician A only
   b. Technician B only
   c. both A and B
   d. neither A nor B

5. Technician A says that spinout failures are due to lubrication throw-off caused by centrifugal force at high rotational speeds. Technician B says that differential carriers increase input speeds by at least a 3:1 ratio. Who is correct?
   a. Technician A only
   b. Technician B only
   c. both A and B
   d. neither A nor B

6. Which of the following gear set types requires the use of an extreme pressure (EP) lubricant?
   a. spiral bevel
   b. crown and pinion
   c. spur bevel
   d. hypoid

7. When backing under a trailer upper coupler, which of the following on the tractor would do the most to prevent a possible spinout?
   a. using the transmission deep reduction gearing
   b. selecting interaxle differential lockout
   c. deflating the trailer air springs
   d. backing under the upper coupler at a high speed

8. Which type of differential carrier would be most likely to have an oil pump to help distribute lubricating oil in the assembly?
   a. one in which hypoid gearing is used
   b. one in which amboid gearing is used
   c. one in which an interaxle differential is used
   d. any double reduction type

9. Which type of axle half shaft only functions to transmit driving torque to the wheel and supports none of the vehicle weight?
   a. semi-floating
   b. full floating
   c. dead axle
   d. steering axle

10. What is the function of a tractor lift axle on a tandem drive axle?
    a. to reduce the over-axle load on three nonlift axles when the unit is fully loaded
    b. to enable the vehicle to corner better when the lift axle is down and fully loaded
    c. to enable the vehicle to corner better when it is up and fully loaded
    d. A and C are correct

JOB SHEET 23.1

Name _____ Station _____ Date _____

Remove a Differential Carrier from a Drive Axle Housing.

Performance Objective(s): Become familiar with the procedure for removing a differential carrier from a drive axle housing.

ASE Education Foundation Correlation

This job sheet addresses the following ASE Education Foundation task(s):

II. Drive Train
  E. Drive Axles
  4. Inspect drive axle shafts; determine needed action. (IMMR, TST, MTST) P-2.
  5. Remove and replace wheel assembly; check rear wheel seal and axle flange for leaks; determine needed action. (IMMR, TST, MTST) P-1.
  9. Remove and replace differential carrier assembly. (TST, MTST) P-2.

Tools and Materials: A truck with a differential carrier, standard shop hand tools, power tools, a transmission jack and carrier adapter, jack stands, prybars, and a differential carrier repair stand

Protective Clothing: Standard shop apparel including coveralls or shop coat, safety glasses, and safety footwear

PROCEDURE
Ensure that the shop LOTO is observed.

1. If the truck is equipped with a dual-range axle, shift the axle to the low range.

   Task completed _____

2. Use a transmission jack and adapter to lift and support the end of the truck where the axle is mounted.

   Task completed _____

3. Place jack stands under each spring seat of the axle to hold the truck in the raised position.

   Task completed _____

4. Remove the plug from the bottom of the axle housing and drain the lubricant from the differential carrier assembly.

   Task completed _____

5. Disconnect the driveline U-joint from the pinion input yoke or flange on the carrier.

   Task completed _____

6. Loosen (without removing) the capscrews and washers or stud nuts and washers from the flanges of both axle shafts.

   Task completed _____

7. Use a brass drift and a large hammer to loosen the tapered dowels in the flanges of both axle shafts.

   Task completed _____

8. Remove the fasteners, tapered dowels, and both axle shafts from the axle assembly.

   Task completed _____

9. On dual-range units disconnect the shift unit air lines. Remove the shift unit, catching oil that will escape from the reservoir.

   Task completed _____

10. Place a transmission jack and adapter under the differential carrier to support the assembly.

    Task completed _____

11. Remove all but the top two carrier-to-housing capscrews or stud nuts and washers.

    Task completed _____

12. Loosen the top two carrier-to-housing fasteners and leave them attached to the assembly.

    Task completed _____

13. Loosen the differential carrier from the axle housing. Use a leather mallet to hit the mounting flange of the carrier at several points.

    Task completed _____

14. Carefully remove the carrier from the axle housing using the transmission jack and adapter. Use a prybar with a radiused end to help remove the carrier from the housing.

    Task completed _____

15. Remove and discard the carrier-to-housing gasket.

    Task completed _____

16. Use a hoist to lift the differential carrier by the input yoke or flange and put the assembly in a repair stand.

    Task completed _____

STUDENT SELF-EVALUATION

Check	Level	Competency	Comments
	4	Mastered task	
	3	Competent but need further help	
	2	Needed a lot of help	
	0	Did not understand the task	

INSTRUCTOR EVALUATION

Check	Level	Competency	Comments
	4	Mastered task	
	3	Competent but needs further help	
	2	Requires more training	
	0	Unable to perform task	

ONLINE TASKS

Use a search engine and research Mack Trucks torque proportioning power dividers. Then answer the following questions:

1. When were these power dividers first used?

2. What are their advantages versus other types of power dividers?

3. What is the cost of a Mack Trucks forward drive axle compared to that of an equivalent rated Eaton?

STUDY TIPS

Identify 5 key points in Chapter 23. Try to be as brief as possible.

Key point 1 _____

Key point 2 _____

Key point 3 _____

Key point 4 _____

Key point 5 _____

24 Heavy-Duty Truck Axle Service and Repair

Objectives

After reading this chapter, you should be able to:

- Describe the lubrication requirements of truck and trailer dead axles.
- Outline the importance of not mixing synthetic- and mineral-based gear lubes.
- Outline the lubrication service procedures required for truck drive axle assemblies.
- Perform some basic level troubleshooting on differential carrier gearing.
- Outline the procedure required to disassemble a differential carrier.
- Disassemble a power divider unit.
- Perform failure analysis on power divider and differential carrier components.
- Reassemble power divider and differential carrier assemblies.

PRACTICE QUESTIONS

1. Technician A says that API GL-5 lubricant is generally recommended for most current heavy-duty truck drive axles. Technician B says that most OEMs recommend oil additives if the objective is to extend service life. Who is correct?
    a. Technician A only
    b. Technician B only
    c. both A and B
    d. neither A nor B

2. What is the recommended gear lube viscosity for differential carriers when operated in severe winter conditions in which temperatures drop to −30°F (−34°C)?
    a. 75W
    b. 80W-90
    c. 80W-140
    d. 85W-140

3. A gear shaft that shears, producing a star-shaped fracture pattern, has most likely been subjected to which of the following overloads?
    a. bending stress
    b. torsional stress
    c. spinout
    d. shock load

4. What is the function of a thrust screw and thrust block?
    a. limits crown gear deflection under load
    b. limits pinion gear deflection under load
    c. defines axle shaft endplay
    d. defines power divider endplay

5. Where is a spigot bearing located?
    a. on each axle shaft
    b. on the differential spider
    c. on the crown gear
    d. on the pinion gear

6. Technician A says that most heavy-duty power dividers can only be removed after the differential carrier assembly has been pulled from the banjo housing. Technician B says that a power divider should measure 0.003–0.007 inch (0.076–0.178 mm) of endplay after installation. Who is correct?
   a. Technician A only
   b. Technician B only
   c. both A and B
   d. neither A nor B

7. Which of the following would be the most likely maximum ring gear runout specification?
   a. 0.002 inch (0.051 mm)
   b. 0.008 inch (0.203 mm)
   c. 0.025 inch (0.635 mm)
   d. 0.030 inch (0.762 mm)

8. Technician A removes ring gear rivets with a hammer and chisel. Technician B drills through the rivet before driving it out with a punch. Who is correct?
   a. Technician A only
   b. Technician B only
   c. both A and B
   d. neither A nor B

9. A differential carrier produces noise on turns only. Which of the following would be the most likely cause?
   a. low lube level
   b. contaminated lube
   c. defective differential pinion gears
   d. defective power divider

10. When checking lube level in a differential carrier, where should the oil level be?
    a. within 1 inch of the fill hole
    b. within one finger of the fill hole
    c. exactly even with the fill hole
    d. slightly above the fill hole

JOB SHEET 24.1

Name _____ Station _____ Date _____

Disassemble a Differential Carrier Assembly.

Performance Objective(s): Using a differential repair stand, disassemble a differential carrier assembly.

ASE Education Foundation Correlation

This job sheet addresses the following ASE Education Foundation task(s):

II. **Drive Train**
   E. **Drive Axles**
   2. Check drive axle fluid level and condition; check drive axle filter; determine needed action. (IMMR, TST, MTST) P-1.
   3. Inspect air-operated power divider (inter-axle differential) assembly including: diaphragms, seals, springs, yokes, pins, lines, hoses, fittings, and controls. (IMMR) P-2.
   3. Inspect and/or adjust air-operated power divider (inter-axle differential) assembly including: diaphragms, seals, springs, yokes, pins, lines, hoses, fittings, and controls. (TST) P-2.
   3. Inspect, adjust, repair, and/or replace air-operated power divider (inter-axle differential) assembly including: diaphragms, seals, springs, yokes, pins, lines, hoses, fittings, and controls. (MTST) P-2.
   6. Inspect, repair, or replace drive axle lubrication system pump, troughs, collectors, slingers, tubes, and filters. (TST, MTST) P-3.
   9. Remove and replace differential carrier assembly. (TST, MTST) P-2.

Tools and Materials: Standard shop hand tools, power tools, a differential carrier overhaul stand, a fish scale, torque wrenches, thread locking compound, and cleaning solvent

Protective Clothing: Standard shop apparel including coveralls or shop coat, safety glasses, and safety footwear

PROCEDURE
Ensure that the shop LOTO is observed.

1. Mount the differential carrier assembly into a test stand.

   Task completed _____

2. If the carrier has a thrust screw, loosen the jam nut on the thrust screw (reference Figure 24–1).

   Task completed _____

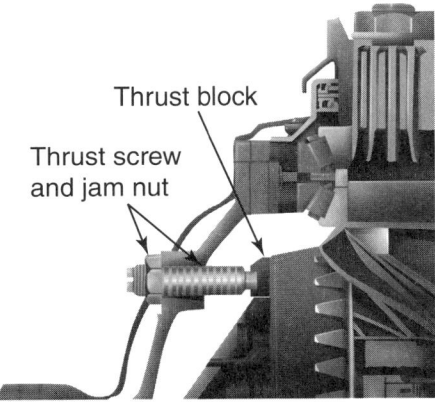

FIGURE 24–1 Thrust screw, jam nut, and thrust block

3. Remove the thrust screw and jam nut from the differential carrier.

Task completed _____

4. Rotate the differential in the repair stand until the ring gear is at the top of the assembly.

Task completed _____

5. Use a center punch and hammer to mark one carrier leg and bearing cap for the purpose of correctly matching the parts when the carrier is reassembled.

Task completed _____

6. Remove the cotter keys, pins, or lock plates that hold the two bearing adjusting rings in position. Use a small drift and hammer to remove the pins.

Task completed _____

7. Remove the capscrews and washers that hold the two bearing caps on the carrier.

Task completed _____

8. Remove the bearing caps and bearing adjusting rings from the carrier.

Task completed _____

9. Lift the differential and ring gear assembly from the carrier and put it on a workbench.

Task completed _____

10. Remove the thrust block (if provided) from inside the carrier.

Task completed _____

STUDENT SELF-EVALUATION

Check	Level	Competency	Comments
	4	Mastered task	
	3	Competent but need further help	
	2	Needed a lot of help	
	0	Did not understand the task	

INSTRUCTOR EVALUATION

Check	Level	Competency	Comments
	4	Mastered task	
	3	Competent but needs further help	
	2	Requires more training	
	0	Unable to perform task	

JOB SHEET 24.2

Name _____ Station _____ Date _____

Assemble a Differential Carrier and Install it into a Drive Axle Housing.

Performance Objective(s): Become familiar with the procedure required to assemble a differential carrier and then install it into a drive axle housing.

ASE Education Foundation Correlation

This job sheet addresses the following ASE Education Foundation task(s):

II. **Drive Train**
   E. **Drive Axles**
   1. Check for fluid leaks; inspect drive axle housing assembly, cover plates, gaskets, seals, vent/breather, and magnetic plugs. (IMMR) P-1.
   1. Check and repair fluid leaks; inspect drive axle housing assembly, cover plates, gaskets, seals, vent/breather, and magnetic plugs. (TST, MTST) P-1.
   2. Check drive axle fluid level and condition; check drive axle filter; determine needed action. (IMMR, TST, MTST) P-1.
   3. Inspect air-operated power divider (inter-axle differential) assembly including: diaphragms, seals, springs, yokes, pins, lines, hoses, fittings, and controls. (IMMR) P-2.
   3. Inspect and/or adjust air-operated power divider (inter-axle differential) assembly including: diaphragms, seals, springs, yokes, pins, lines, hoses, fittings, and controls. (TST) P-2.
   3. Inspect, adjust, repair, and/or replace air-operated power divider (inter-axle differential) assembly including: diaphragms, seals, springs, yokes, pins, lines, hoses, fittings, and controls. (MTST) P-2.
   4. Inspect drive axle shafts; determine needed action. (IMMR, TST, MTST) P-2.
   5. Remove and replace wheel assembly; check rear wheel seal and axle flange for leaks; determine needed action. (IMMR, TST, MTST) P-1.
   6. Inspect, repair, or replace drive axle lubrication system pump, troughs, collectors, slingers, tubes, and filters. (TST, MTST) P-3.
   9. Remove and replace differential carrier assembly. (TST, MTST) P-2.

Tools and Materials: A differential carrier assembly, standard shop hand and power tools, a differential carrier repair stand, a fish scale, torque wrenches, thread locking compound, and cleaning solvent

Protective Clothing: Standard shop apparel including coveralls or shop coat, safety glasses, and safety footwear

PROCEDURE
Ensure that the shop LOTO is observed.

1. Clean and dry the bearing cups and bores of the carrier legs and bearing caps.

   Task completed _____

2. Apply axle lubricant on the inner diameter of the bearing cups and onto both bearing cones that are assembled on the case halves.

   Task completed _____

3. Apply thread locking compound in the bearing bores of the carrier legs and bearing caps.

   Task completed _____

4. Install the bearing cups over the bearing cones that are assembled on the case halves.

   Task completed _____

5. Lift the differential and ring gear assembly and install it into the carrier.

    Task completed _____

6. Install both the bearing adjusting rings into position between the carrier legs. Turn each adjusting ring hand-tight against the bearing cup.

    Task completed _____

7. Install the bearing caps over the bearings and adjusting rings in the correct location as marked before removal.

    Task completed _____

8. Tap each bearing cap into position with a light leather, plastic, or rubber mallet.

    Task completed _____

9. Install the hardware that holds the bearing caps to the carrier. Tighten all hardware by hand first and then torque to the correct values.

    Task completed _____

10. Using a fish scale or light-duty torque wrench, adjust the preload of the differential bearing.

    Task completed _____

11. Check the runout of the ring gear using a dial indicator.

    Task completed _____

12. Adjust the backlash of the ring gear.

    Task completed _____

13. Check and adjust the tooth contact pattern using machinists' blueing. Figure 24–2 shows a gear tooth contact pattern.

    Task completed _____

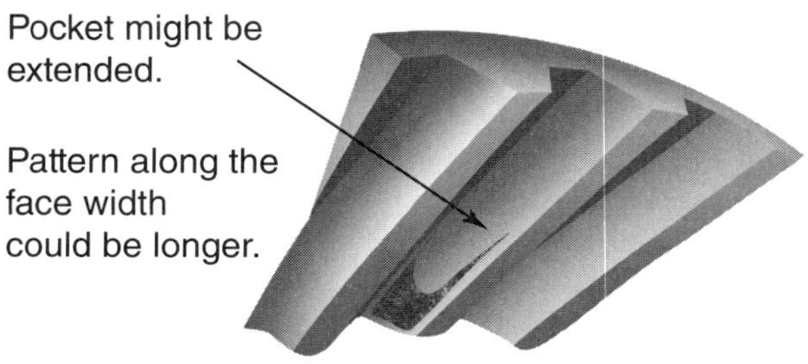

Pocket might be extended.

Pattern along the face width could be longer.

FIGURE 24–2 Gear tooth contact pattern

14. Adjust the thrust screw to the manufacturer specification.

    Task completed _____

15. Install the cotter keys or lock plates that hold the bearing adjusting rings in position.

    Task completed _____

16. With a cleaning solvent and clean shop cloths, clean the inside of the axle housing and the mounting surface where the carrier fastens.

    Task completed _____

17. Blow-dry the cleaned areas.

    Task completed _____

18. Inspect the axle housing for damage; repair or replace the axle housing if damaged.

    Task completed _____

19. Check for loose studs in the mounting surface of the housing where the carrier fastens.

    Task completed _____

20. Remove and clean the studs that are loose.

    Task completed _____

21. Apply locking compound to the threaded holes and torque the studs into the axle housing.

    Task completed _____

22. Apply silicone gasket material to the mounting face of the banjo housing.

    Task completed _____

23. Install fasteners or guide studs in the four corner locations around the carrier and axle housing. Tighten the fasteners hand-tight at this time.

    Task completed _____

24. Gently push the carrier into position so that the carrier and banjo housing faces contact.

    Task completed _____

25. Tighten the four fasteners two or three turns each in a pattern opposite each other.

    Task completed _____

26. Torque the four fasteners to the correct value.

    Task completed _____

27. Install the remaining fasteners that hold the carrier in the axle housing. Tighten to the correct torque value.

    Task completed _____

28. Connect the driveline U-joint to the pinion input yoke or flange on the carrier.

    Task completed _____

29. Install new gaskets and axle shafts into the axle housing and carrier. The gasket and flange of the axle shafts must fit flat against the wheel hub.

    Task completed _____

30. Install the axle shaft fasteners to the wheel hubs. Tighten all fasteners to the correct torque value.

    Task completed _____

31. Install the tapered dowels at each stud and into the flange of the axle shaft.

Task completed _____

32. Install the hardware on the studs and tighten to the correct torque value.

Task completed _____

STUDENT SELF-EVALUATION

Check	Level	Competency	Comments
	4	Mastered task	
	3	Competent but need further help	
	2	Needed a lot of help	
	0	Did not understand the task	

INSTRUCTOR EVALUATION

Check	Level	Competency	Comments
	4	Mastered task	
	3	Competent but needs further help	
	2	Requires more training	
	0	Unable to perform task	

ONLINE TASKS

Identify a specific truck differential carrier by serial number and then use a search engine and research the following:

1. Cost of rebuilding the differential carrier when the differential gearing and all bearings have to be replaced
2. Cost of a rebuilt exchange unit minus core charge

STUDY TIPS

Identify 5 key points in Chapter 24. Try to be as brief as possible.

Key point 1 _____

Key point 2 _____

Key point 3 _____

Key point 4 _____

Key point 5 _____

25 Steering and Alignment

Objectives

After reading this chapter, you should be able to:

- Identify the components of the steering system of a heavy-duty truck.
- Describe the procedure for inspecting front axle components for wear.
- Explain how toe, camber, caster, axle inclination, turning radius, and axle alignment affect tire wear, directional stability, and handling.
- Describe the components and operation of a worm and sector shaft and a recirculating ball-type steering gear.
- Explain how to check and adjust a manual steering gear preload and backlash.
- Identify the components of a power steering gear and pump and explain the operation of a power steering system.
- Outline the operating principles of a truck rack and pinion steering gear.
- Describe the components and operation of an electronically variable power steering system.
- Explain how electronically managed steering is a key to enabling semi- and fully autonomous truck operation.
- Describe the components and operation of a load-sensing power-assist steering system.

PRACTICE QUESTIONS

1. Technician A says that a shorter Pitman arm will produce more steering angle motion at the front wheels for a given amount of steering wheel movement. Technician B says that an advantage of a two-piece drag link is that it is adjustable. Who is correct?
   a. Technician A only
   b. Technician B only
   c. both A and B
   d. neither A nor B

2. Which of the following front-end geometry factors describes the tracking of each front wheel in relation to the other?
   a. toe
   b. caster
   c. camber
   d. KPI

3. Technician A says that too little caster can cause steering wander and poor recovery. Technician B says that incorrect toe angles cause more front-end wear than all the other front alignment angles. Who is correct?
   a. Technician A only
   b. Technician B only
   c. both A and B
   d. neither A nor B

4. Which of the following describes the inward or outward tilt at the top of the wheels when observed from the front of the vehicle?
   a. toe
   b. caster
   c. camber
   d. none of the above

5. Technician A says that steering axle components should be lubricated every 25,000 miles (38,625 km). Technician B says that front wheel bearings should be inspected and lubricated every time the wheel hub is removed. Who is correct?
   a. Technician A only
   b. Technician B only
   c. both A and B
   d. neither A nor B

6. Technician A says that when the toe adjustment is out of specification, leading and darting can occur. Technician B says that when steering gear total mesh preload is low, shimmy can occur at highway speeds. Who is correct?
   a. Technician A only
   b. Technician B only
   c. both A and B
   d. neither A nor B

7. Technician A says that manual steering gear would be better suited to a truck in a linehaul application than in a pickup and delivery application, because steering effort required is greatest at low-speed maneuvering. Technician B says that all hydraulic power-assisted steering in current trucks must default to manual steering in the event of hydraulic failure. Who is correct?
   a. Technician A only
   b. Technician B only
   c. both A and B
   d. neither A nor B

8. Technician A says that the need for hydraulic-assist in a hydraulic steering gear increases in proportion to the road speed of the vehicle. Technician B says that the highest hydraulic-assist pressures can be measured when the vehicle is fully loaded and stationary. Who is correct?
   a. Technician A only
   b. Technician B only
   c. both A and B
   d. neither A nor B

9. A driver reports that his vehicle is producing hard steering. In what sequence should the following checks be performed?
   a. power steering flow and pressure test
   b. kingpins
   c. fifth wheel lubrication
   d. front-end geometry

10. What is the most commonly used size for a heavy-duty truck steering wheel?
    a. 15 inch diameter (380 mm)
    b. 22 inch diameter (560 mm)
    c. 25 inch diameter (635 mm)
    d. 28 inch diameter (710 mm)

JOB SHEET 25.1

Name _____ Station _____ Date _____

Measure Caster.

Performance Objective(s): Use a machinist's protractor to measure caster on a truck chassis and compare to OEM specifications.

ASE Education Foundation Correlation

This job sheet addresses the following ASE Education Foundation task(s):

IV. Suspension and Steering Systems
   A. General
   1. Research vehicle service information, including fluid type, vehicle service history, service precautions, and technical service bulletins. (IMMR, TST, MTST) P-1.

   F. Wheel Alignment
   1. Demonstrate understanding of alignment angles. (IMMR) P-3.

   F. Wheel Alignment Diagnosis and Repair
   1. Demonstrate understanding of alignment angles. (TST, MTST) P-1.
   4. Check and record caster. (TST, MTST) P-2.

Tools and Materials: A truck chassis, a machinist's protractor, and standard shop and hand tools

Protective Clothing: Standard shop apparel including coveralls or shop coat, safety glasses, and safety footwear

PROCEDURE
Ensure that the shop LOTO is observed.

1. Place the vehicle on a level shop floor. Drive back and forth a couple of times with the steering wheel straight ahead to ensure that there are no steering components under side load.

   Task completed _____

2. Check that the frame is at its specified height both at the front and back. On units with air suspension at the rear, ensure that this is adjusted to the correct height.

   Task completed _____

3. Clean any dirt and debris from the axle pads.

   Task completed _____

4. Place the machinist's protractor on the machined surface of the front axle pad and ensure that it contacts both spring U-bolts; that is, it should be exactly parallel with the truck frame.

   Task completed _____

5. Center the bubble in the protractor level by rotating the protractor dial. Lock the dial, remove the protractor, and record the reading.

   Task completed _____

6. Determine whether the reading is positive or negative and compare it to specification.

   Task completed _____

7. Repeat the preceding sequence on the other side of the vehicle, again recording the camber angle.

   Task completed _____

8. A small camber angle differential between the two sides may be specified by the OEM. Check the OEM specifications before making an adjustment.

   Task completed _____

STUDENT SELF-EVALUATION

Check	Level	Competency	Comments
	4	Mastered task	
	3	Competent but need further help	
	2	Needed a lot of help	
	0	Did not understand the task	

INSTRUCTOR EVALUATION

Check	Level	Competency	Comments
	4	Mastered task	
	3	Competent but needs further help	
	2	Requires more training	
	0	Unable to perform task	

JOB SHEET 25.2

Name _____ Station _____ Date _____

Adjust Caster Angle.

Performance Objective(s): Become familiar with the procedure to adjust caster angle to specification using a machinist's square and standard shop tools.

ASE Education Foundation Correlation

This job sheet addresses the following ASE Education Foundation task(s):

IV. Suspension and Steering Systems
   A. General
   1. Research vehicle service information, including fluid type, vehicle service history, service precautions, and technical service bulletins. (IMMR, TST, MTST) P-1.

   F. Wheel Alignment
   1. Demonstrate understanding of alignment angles. (IMMR) P-3.

   F. Wheel Alignment Diagnosis and Repair
   1. Demonstrate understanding of alignment angles. (TST, MTST) P-1.
   4. Check and record caster. (TST, MTST) P-2.

Tools and Materials: A truck with a front-end caster angle that has been measured out of specification and a frame jack

Protective Clothing: Standard shop apparel including coveralls or shop coat, safety glasses, and safety footwear

PROCEDURE
Ensure that the shop LOTO is observed.

1. Measure the caster angle using the procedure used in Job Sheet Number 25.1.

   Task completed _____

2. Loosen the front spring U-bolts on each side of the vehicle. Then, on one side only, completely remove the U-bolts on one side of the vehicle.

   Task completed _____

3. Use a frame jack to hoist the frame and lift the spring off the axle seat. Do not lift so much that the wheel begins to lift off the floor.

   Task completed _____

4. If a caster shim has been previously installed, remove this and check its angle. This will have to be factored when calculating the desired caster angle.

   Task completed _____

5. Use a wire brush and clean the axle pad thoroughly.

   Task completed _____

6. Check the new caster shim and ensure that the spring center bolt head will protrude through it so it can engage into the axle recess.

   Task completed _____

7. Insert the shim and carefully drop the springs into the caster shim and axle by lowering the vehicle. Make sure that the spring pin head is fully engaged into the axle recess.

Task completed _____

8. Install the nuts onto the spring U-bolts and torque to specification.

Task completed _____

9. Repeat the previous procedure on the other side of the vehicle.

Task completed _____

10. Use Job Sheet Number 25.1 to check that the adjusted caster angle meets the OEM specification.

Task completed _____

STUDENT SELF-EVALUATION

Check	Level	Competency	Comments
	4	Mastered task	
	3	Competent but need further help	
	2	Needed a lot of help	
	0	Did not understand the task	

INSTRUCTOR EVALUATION

Check	Level	Competency	Comments
	4	Mastered task	
	3	Competent but needs further help	
	2	Requires more training	
	0	Unable to perform task	

JOB SHEET 25.3

Name _____ Station _____ Date _____

Measure and Adjust Front Axle Toe Settings.

Performance Objective(s): Use a tram bar, scriber, and chalk to measure front axle toe and make an adjustment if necessary.

ASE Education Foundation Correlation

This job sheet addresses the following ASE Education Foundation task(s):

IV. Suspension and Steering Systems
  A. General
  1. Research vehicle service information, including fluid type, vehicle service history, service precautions, and technical service bulletins. (IMMR, TST, MTST) P-1.
  F. Wheel Alignment
  1. Demonstrate understanding of alignment angles. (IMMR) P-3.
  F. Wheel Alignment Diagnosis and Repair
  1. Demonstrate understanding of alignment angles. (TST, MTST) P-1.
  5. Check, record, and adjust toe settings. (TST, MTST) P-1.
  8. Check front axle alignment (centerline). (TST, MTST) P-2.

Tools and Materials: A truck with single axle steering, a tram bar, a tire scriber (disc type), chalk, and standard shop hand and power tools

Protective Clothing: Standard shop apparel including coveralls or shop coat, safety glasses, and safety footwear

PROCEDURE
Ensure that the shop LOTO is observed.

1. Jack up the truck front end and rotate each tire, applying chalk dust to a flat tire land while rotating the wheel through one revolution. The practice of using spray paint instead of chalk is not recommended, because the chemicals in the paint can cause surface damage to the tire rubber.

   Task completed _____

2. Mount a disk-type tire scriber on a jack stand and make the shallowest possible scribe mark on the periphery of each tire by rotating it through one revolution.

   Task completed _____

3. Lower the front end of the truck back on the floor.

   Task completed _____

4. Align the fixed tram bar pointer to the scribed line on one tire from the rear. Use the adjustable pointer on the tram bar to align it with the scribed line on the tire on the opposite side. Lock the adjustable pointer into position and remove the tram bar from under the truck.

   Task completed _____

5. Now place the tram bar in front of the steering tires. Align one of the pointers with the scribed line on one of the tires. Check the opposite side. If this pointer also perfectly aligns with the scribed line on the tire, there is zero toe. If the pointer registers outside the scribed line, there is a toe-in factor, and this can be measured. If the pointer registers inside the scribed line, there is a toe-out factor. It is essential to perform this measurement with the full weight of the truck on the wheels.

   Task completed _____

6. Check the toe specification with the OEM's required specifications. This may change with the type of tires used. Specifications for highway vehicles equipped with radial tires are typically zero toe, and if equipped with bias ply tires, between 1/16 and 1/8 inch (1.5 and 3 mm) of toe-in.

   Task completed _____

7. To adjust toe, raise the vehicle so that the front wheels are off the floor and place it on jack stands. Loosen the toe bar clamps. Rotate the toe bar (threaded to the toe ends) to either increase or decrease the toe bar length. Drop the unit back on the floor to perform a toe measurement after the adjustment.

   Task completed _____

8. When a toe adjustment has been completed, ensure that the toe bar clamps are torqued to specification.

   Task completed _____

STUDENT SELF-EVALUATION

Check	Level	Competency	Comments
	4	Mastered task	
	3	Competent but need further help	
	2	Needed a lot of help	
	0	Did not understand the task	

INSTRUCTOR EVALUATION

Check	Level	Competency	Comments
	4	Mastered task	
	3	Competent but needs further help	
	2	Requires more training	
	0	Unable to perform task	

JOB SHEET 25.4

Name _____ Station _____ Date _____

Analyze a Hydraulic Power-Assisted Steering System.

Performance Objective(s): Use a set of power steering pressure and flow gauges to analyze the performance of a hydraulic power-assisted steering system in a truck.

ASE Education Foundation Correlation

This job sheet addresses the following ASE Education Foundation task(s):

IV. Suspension and Steering Systems
A. General
1. Research vehicle service information, including fluid type, vehicle service history, service precautions, and technical service bulletins. (IMMR, TST, MTST) P-1.

C. Steering Pump and Gear Units
1. Check power steering pump and gear operation, mountings, lines, and hoses; check fluid level and condition; service filter; inspect system for leaks. (IMMR, TST, MTST) P-1.
2. Flush and refill power steering system; purge air from system. (IMMR, TST, MTST) P-2.
3. Identify causes of power steering system noise, binding, darting/oversteer, reduced wheel cut, steering wheel kick, pulling, non recovery, turning effort, looseness, hard steering, overheating, fluid leakage, and fluid aeration problems. (TST) P-1.
3. Diagnose causes of power steering system noise, binding, darting/oversteer, reduced wheel cut, steering wheel kick, pulling, non recovery, turning effort, looseness, hard steering, overheating, fluid leakage, and fluid aeration problems. (MTST) P-1.
4. Inspect, service, and/or replace power steering reservoir, seals, and gaskets. (TST, MTST) P-2.
6. Inspect and/or replace power steering gear(s) (single and/or dual) and mountings. (TST, MTST) P-2.

Tools and Materials: A truck with hydraulic power-assisted steering, pressure and flow gauges, a flow control valve, a catch basin, and standard shop and power tools

Protective Clothing: Standard shop apparel including coveralls or shop coat, safety glasses, and safety footwear

PROCEDURE
Ensure that the shop LOTO is observed.

1. With the engine off, disconnect the pressure line feeding fluid from the power steering pump to the steering gear. Place a drain pan below the coupler to catch any oil that might spill. Separate at the steering gear.

   Task completed _____

2. Verify the type of hydraulic fluid (engine oil, auto transmission fluid, etc.).

   Task completed _____

3. Connect the power steering analyzer in series into the pump-to-steering gear circuit, making sure that the gauges can be read and the flow control valve accessed during the testing.

   Task completed _____

4. Ensure that the flow control valve is fully open.

   Task completed _____

5. Start the engine and, running it at idle, work the steering wheel from full right lock to full left lock several times to ensure that air is purged from the circuit.

   Task completed _____

6. Choke down on the flow control valve and note the flow and pressure values. Compare the peak flow and pressure values to the OEM specification.

   Task completed _____

7. Accelerate the engine to 1,500 rpm. Perform the pressure and flow tests and compare the data measured to the OEM specifications.

   Task completed _____

8. Note the system relief valve setting.

   Task completed _____

9. Remove the power steering analyzer gauges from the vehicle, remembering that the fluid has probably been heated up during testing.

   Task completed _____

10. Check the vehicle power steering reservoir fluid level and top off if required.

    Task completed _____

STUDENT SELF-EVALUATION

Check	Level	Competency	Comments
	4	Mastered task	
	3	Competent but need further help	
	2	Needed a lot of help	
	0	Did not understand the task	

INSTRUCTOR EVALUATION

Check	Level	Competency	Comments
	4	Mastered task	
	3	Competent but needs further help	
	2	Requires more training	
	0	Unable to perform task	

JOB SHEET 25.5

Name _____ Station _____ Date _____

Poppet Adjustment on TAS and THP Steering Gear.

Performance Objective(s): Set the poppets and axle stops on TRW TAS and THP series steering gear after installation.

ASE Education Foundation Correlation

This job sheet addresses the following ASE Education Foundation task(s):

IV. Suspension and Steering Systems
   A. General
   1. Research vehicle service information, including fluid type, vehicle service history, service precautions, and technical service bulletins. (IMMR, TST, MTST) P-1.

   C. Steering Pump and Gear Units
   1. Check power steering pump and gear operation, mountings, lines, and hoses; check fluid level and condition; service filter; inspect system for leaks. (IMMR, TST, MTST) P-1.
   3. Identify causes of power steering system noise, binding, darting/oversteer, reduced wheel cut, steering wheel kick, pulling, non recovery, turning effort, looseness, hard steering, overheating, fluid leakage, and fluid aeration problems. (TST) P-1.
   3. Diagnose causes of power steering system noise, binding, darting/oversteer, reduced wheel cut, steering wheel kick, pulling, non recovery, turning effort, looseness, hard steering, overheating, fluid leakage, and fluid aeration problems. (MTST) P-1.
   4. Inspect, service, and/or replace power steering reservoir, seals, and gaskets. (TST, MTST) P-2.

   F. Wheel Alignment Diagnosis and Repair
   7. Identify turning/Ackerman angle (toe-out-on-turns) problems. (TST, MTST) P-3.

Tools and Materials: A truck with hydraulic power-assisted TAS37, TAS40, TAS55, TAS65, TAS85, THP45, or THP60 steering gear—this generation of steering gear uses automatic poppet valve setting; standard shop and power tools

Protective Clothing: Standard shop apparel including coveralls or shop coat, safety glasses, and safety footwear

PROCEDURE
Ensure that the shop LOTO is observed.

1. Install the steering gear.

   Task completed _____

2. Plumb in the steering gear pressure and return lines.

   Task completed _____

3. If a slave gear is installed (dual steering axles), plumb in the lines to the slave gear.

   Task completed _____

4. Install the Pitman arm onto the steering gear sector shaft. Ensure that the timing marks on the arm and shaft are aligned.

   Task completed _____

5. Install the steering column yoke and secure to specification.

   Task completed _____

6. Check the power steering fluid level and top up if required.

   Task completed _____

7. Raise the front steering axle off the shop floor.

   Task completed _____

8. Start the engine. Turn the steering wheel from left lock to right lock two times to bleed air from the hydraulic circuit. Center the steer axle wheels. Shut the engine off and check the power steering fluid level and correct if required.

   Task completed _____

9. Lower the front axle. Start the engine. Once again turn the steering wheel to full right lock and to full left lock. This may require some effort. This process sets the automatic poppets on the steering gear.

   Task completed _____

STUDENT SELF-EVALUATION

Check	Level	Competency	Comments
	4	Mastered task	
	3	Competent but need further help	
	2	Needed a lot of help	
	0	Did not understand the task	

INSTRUCTOR EVALUATION

Check	Level	Competency	Comments
	4	Mastered task	
	3	Competent but needs further help	
	2	Requires more training	
	0	Unable to perform task	

JOB SHEET 25.6

Name _____ Station _____ Date _____

Set the Axle Stops on an M-Series Steering Gear.

Performance Objective(s): Set the axle stops on an M-Series steering gear after installation.

ASE Education Foundation Correlation

This job sheet addresses the following ASE Education Foundation task(s):

IV. Suspension and Steering Systems
   A. General
   1. Research vehicle service information, including fluid type, vehicle service history, service precautions, and technical service bulletins. (IMMR, TST, MTST) P-1.

   C. Steering Pump and Gear Units
   1. Check power steering pump and gear operation, mountings, lines, and hoses; check fluid level and condition; service filter; inspect system for leaks. (IMMR, TST, MTST) P-1.
   3. Identify causes of power steering system noise, binding, darting/oversteer, reduced wheel cut, steering wheel kick, pulling, non recovery, turning effort, looseness, hard steering, overheating, fluid leakage, and fluid aeration problems. (TST) P-1.
   3. Diagnose causes of power steering system noise, binding, darting/oversteer, reduced wheel cut, steering wheel kick, pulling, non recovery, turning effort, looseness, hard steering, overheating, fluid leakage, and fluid aeration problems. (MTST) P-1.
   4. Inspect, service, and/or replace power steering reservoir, seals, and gaskets. (TST, MTST) P-2.

   F. Wheel Alignment Diagnosis and Repair
   7. Identify turning/Ackerman angle (toe-out-on-turns) problems. (TST, MTST) P-3.

Tools and Materials: A truck with hydraulic power-assisted M90, M100, M110 Series 1 steering gear; standard shop and power tools

Protective Clothing: Standard shop apparel including coveralls or shop coat, safety glasses, and safety footwear

PROCEDURE
Ensure that the shop LOTO is observed.

CAUTION! Do not use this procedure to adjust M-Series 3 steering gear

1. Install the steering gear.

   Task completed _____

2. Plumb in the steering gear pressure and return lines.

   Task completed _____

3. Fill the power steering reservoir to the full mark.

   Task completed _____

4. Install the Pitman arm onto the steering gear sector shaft. Ensure that the timing marks on the arm and shaft are aligned.

   Task completed _____

5. Install the steering column yoke and secure to specification.

   Task completed _____

6. Start the engine. Check the power steering fluid level with the engine running and top up if required.

   Task completed _____

7. Turn the steering wheel to full lock on the left.

   Task completed _____

8. Identify the left relief plunger (located in a ¼-inch (6 mm) hole in between two of the four bolts that hold the bearing cover on the head) on the steering gear. Have someone hold the steering wheel onto full left lock.

Task completed _____

9. Using a screwdriver, back off the plunger (turn CCW) so that the wheel fully contacts the stop. This adjusts the left stop. Repeat the process for the right stop. The steering gear plunger for the right stop is located on the opposite end of the steering gear.

Task completed _____

CAUTION! Do not turn the plunger out past the surface of the head because it may blow out with extreme force.

STUDENT SELF-EVALUATION

Check	Level	Competency	Comments
	4	Mastered task	
	3	Competent but need further help	
	2	Needed a lot of help	
	0	Did not understand the task	

INSTRUCTOR EVALUATION

Check	Level	Competency	Comments
	4	Mastered task	
	3	Competent but needs further help	
	2	Requires more training	
	0	Unable to perform task	

ONLINE TASKS

You want to spec out a steering gear for a heavy-dump truck. Get online and use search engines to make your selection. Following are some clues:

1. Shephard
2. Ross TRW
3. HFB-70

STUDY TIPS

Identify 5 key points in Chapter 25. Try to be as brief as possible.

Key point 1 _____

Key point 2 _____

Key point 3 _____

Key point 4 _____

Key point 5 _____

26 Suspension Systems

Objectives

After reading this chapter, you should be able to:

- Identify the suspension systems used on current trucks.
- Describe the components used on mechanical leaf and multileaf spring suspension systems and explain how they work.
- Describe a fiber composite spring.
- Identify equalizing beam suspension system components and explain how they function.
- Describe how rubber spring and torsion bar suspensions function.
- Identify air spring suspension system components and explain how they function.
- Outline the advantages of combining different spring types in a truck suspension system.
- Troubleshoot suspensions and locate defective suspension system components.
- Outline suspension system repair and replacement procedures.
- Explain the relationship between axle alignment and suspension system alignment.
- Troubleshoot common suspension problems.
- Outline some common suspension repair procedures.
- Perform full chassis suspension system alignments.
- Describe the operation of the cab air suspension system.

PRACTICE QUESTIONS

1. The main leaf in a multileaf spring pack has fractured. Technician A says that the spring can be safely rebuilt by replacing the main leaf and reusing the old spring leaves. Technician B says that anytime a spring leaf fails, the whole spring pack should be replaced. Who is correct?
    a. Technician A only
    b. Technician B only
    c. both A and B
    d. neither A nor B

2. The bushed component through which the spring pin is inserted is known as the:
    a. hanger
    b. hook
    c. eye
    d. shackle

3. What is the advantage of a spring design consisting of a single tapered leaf?
    a. higher load capacity
    b. lighter overall weight
    c. better rebound characteristics
    d. variable spring rate

4. The rubber bag on an air spring is mounted on a(n):
   a. pad
   b. pedestal
   c. equalizer
   d. cross-member
5. The solid rubber block mounted on the pedestal inside an air spring assembly is called a:
   a. jounce block
   b. rebound chock
   c. deflection block
   d. riser block
6. During disassembly, if all the jounce blocks inside the air springs show evidence of being hammered out, which of the following would be a likely cause?
   a. worn-out shock absorbers
   b. high tire inflation pressures
   c. overloading
   d. aggressive braking
7. What is the advantage of a lift axle assembly located ahead of the tandem drive axles on a dump truck?
   a. increased loads
   b. better load-over-axle distribution
   c. no tire scuffing when raised during turns
   d. all of the above
8. Which of the following best describes the function of a shock absorber?
   a. dampens spring rebound-to-jounce oscillations
   b. dampens axle longitudinal forces
   c. reduces axle wind-up during acceleration
   d. reduces axle wind-up during braking
9. What gives a multileaf spring assembly its self-dampening characteristics?
   a. a lower torsion bar
   b. friction between the individual leaves
   c. the rebound clips
   d. U-bolt tension
10. Which of the following suspensions would require the use of shock absorbers?
    a. multileaf spring
    b. air spring
    c. air and multileaf spring
    d. solid rubber block

JOB SHEET 26.1

Name _____ Station _____ Date _____

Replace the Equalizer Assembly.

Performance Objective(s): Become familiar with the process required to replace an equalizer assembly in a heavy-duty truck or trailer.

ASE Education Foundation Correlation

This job sheet addresses the following ASE Education Foundation task(s):

IV. Suspension and Steering Systems
  A. General
  1. Research vehicle service information, including fluid type, vehicle service history, service precautions, and technical service bulletins. (IMMR, TST, MTST) P-1.
  E. Suspension Systems
  1. Inspect shock absorbers, bushings, brackets, and mounts; determine needed action. (IMMR) P-1.
  1. Inspect, service, repair, and/or replace shock absorbers, bushings, brackets, and mounts. (TST, MTST) P-1.
  2. Inspect leaf springs, center bolts, clips, pins, bushings, shackles, U-bolts, insulators, brackets, and mounts; determine needed action. (IMMR) P-1.
  2. Inspect, repair, and/or replace leaf springs, center bolts, clips, pins, bushings, shackles, U-bolts, insulators, brackets, and mounts; determine needed action. (TST) P-1.
  2. Inspect, repair, and/or replace leaf springs, center bolts, clips, pins, bushings, shackles, U-bolts, insulators, brackets, and mounts. (MTST) P-1.
  3. Inspect axle and axle aligning devices such as: radius rods, track bars, stabilizer bars, and torque arms; inspect related bushings, mounts, and shims. (IMMR) P-1.
  3. Inspect, repair, and/or replace axle and axle aligning devices such as: radius rods, track bars, stabilizer bars, and torque arms; inspect related bushings, mounts, shims and attaching hardware; determine needed action. (TST, MTST) P-1.
  4. Inspect tandem suspension equalizer components. (IMMR) P-3.
  4. Inspect, repair, and/or replace tandem suspension equalizer components; determine needed action. (TST, MTST) P-1.
  8. Measure, record and adjust ride height; determine needed action. (TST, MTST) P-1.

Tools and Materials: A heavy-duty truck or trailer equipped with an equalizer-type suspension system, standard shop jacking and hoisting equipment, and hand and power tools

Protective Clothing: Standard shop apparel including coveralls or shop coat, safety glasses, and safety footwear

PROCEDURE
Ensure that the shop LOTO is observed.

Reference Figure 26–1, which shows a typical equalizer suspension.

1. Park the truck on level ground and ensure that there is no lateral tension on the suspension components by moving the vehicle forward and backward before parking. Block both sides of the vehicle's front tires.

   Task completed _____

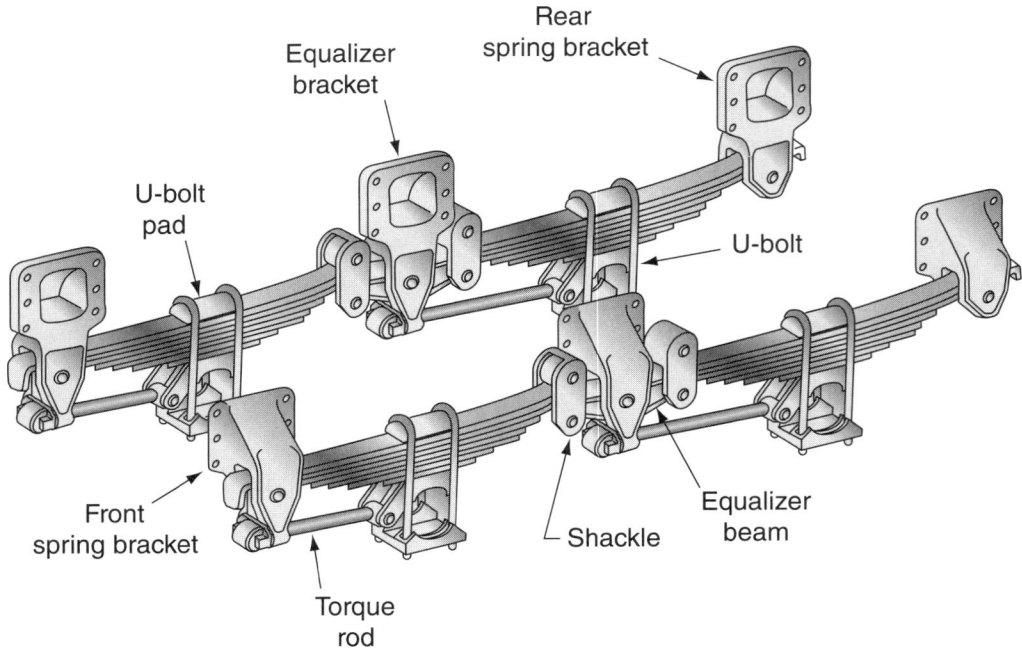

FIGURE 26–1 Equalizer suspension

2. Raise the rear of the truck and support the axles with safety stands.

   Task completed _____

3. Raise the truck frame so that most of the weight is removed from the leaf springs. Now support the frame in its raised position with safety stands.

   Task completed _____

4. Remove the wheel assemblies on that side of the vehicle where the equalizer is to be replaced.

   Task completed _____

5. If removing an equalizer from a truck with tandem drive axles (reference Figure 26–2), follow steps a and b.
   a. Remove the cotter pin from the outboard end of each spring retainer pin.

   Task completed _____

   b. Remove the retainer pins.

   Task completed _____

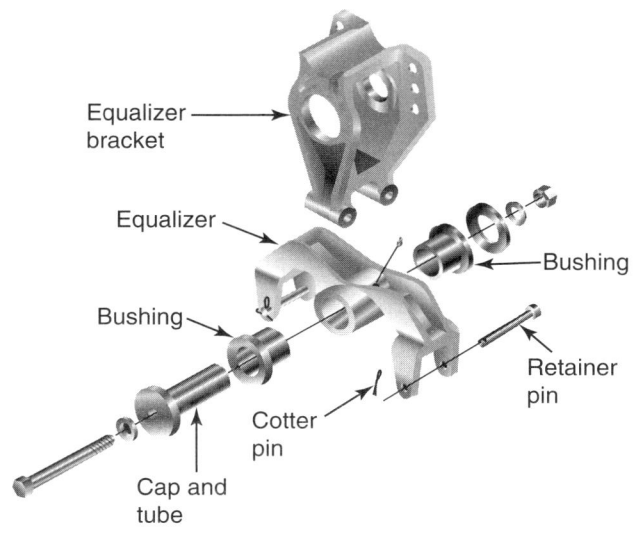
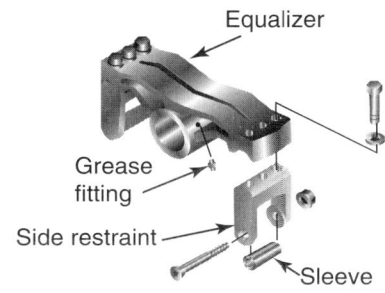

FIGURE 26–2 Equalizer assembly replacement

6. If removing an equalizer from a truck with a pusher or tag axle, follow steps a through c.

   a. Remove the nuts from the flathead bolts in the wear shoe side restraints on each end of the equalizer.

   Task completed _____

   b. Remove the flathead bolts and side restraint sleeves.

   Task completed _____

   c. Remove the capscrews and washers and both wear shoe side restraints from the equalizer.

   Task completed _____

7. Remove the equalizer cap and tube assembly lockout, inboard bearing washer, bolt, and outboard bearing washer.

   Task completed _____

8. Insert a bar between the bottom of the equalizer cap and tube assembly.

   Task completed _____

9. Gently lever the weight of the equalizer off the equalizer cap and tube assembly.

   Task completed _____

10. Insert a piece of bar stock through the inboard equalizer cap and tube assembly bolt hole, and lightly tap the cap and tube assembly out of the equalizer.

    Task completed _____

11. Remove the equalizer from the equalizer bracket.

    Task completed _____

12. Remove the wear washer(s) and equalizer bushings from the equalizer.

    Task completed _____

13. Thoroughly clean the equalizer bushings and inspect them for wear, damage, or defects.

    Task completed _____

14. Replace the bushings if necessary.

    Task completed _____

15. Apply chassis grease to the inside of the equalizer bushings.

    Task completed _____

16. Install the bushings in the equalizer.

    Task completed _____

17. Install the new equalizer in the equalizer bracket.

    Task completed _____

18. Apply chassis grease to the equalizer cap and tube assembly.

    Task completed _____

19. Start the cap and tube assembly into the equalizer through the equalizer bracket.

    Task completed _____

20. Push the equalizer cap and tube assembly part of the way through the equalizer, and then place the inboard wear washer(s) between the inboard equalizer bushing and the equalizer bracket.

    Task completed _____

21. Push the cap and tube assembly the rest of the way into the equalizer bracket.

    Task completed _____

22. Place the outboard bearing washer on the equalizer cap and tube assembly bolt.

    Task completed _____

23. Install the inboard bearing washer and locknut on the cap and tube assembly bolt.

    Task completed _____

24. Tighten the locknut to the torque value specified by the manufacturer.

    Task completed _____

25. If installing an equalizer on a truck with tandem drive axles, follow steps a through d.

    Not applicable _____

    Task completed _____

    a. Apply sealant to the spring retainer pins.

    Task completed _____

    b. Install the pins from the inboard side.

    Task completed _____

c. Ensure that the ends of the spring leaves are above the retainer pins.

   Task completed _____

   d. Install a cotter pin in the outboard end of each retainer pin and lock it in place.

   Task completed _____

26. If installing an equalizer on a truck with a pusher or tag axle, follow steps a through f.

    Not applicable _____

    Task completed _____

    a. Apply sealant to the surfaces where the wear shoe side restraints contact the equalizer.

    Task completed _____

    b. Attach the side restraints to the equalizer, offsetting them toward the inboard side of the equalizer.

    Task completed _____

    c. Tighten and torque the equalizer wear shoe capscrews.

    Task completed _____

    d. Install the side restraint sleeves and flathead bolts in the wear shoe side restraints.

    Task completed _____

    e. Be sure the ends of the spring leaves are above the side restraint sleeves.

    Task completed _____

    f. Install the nuts and tighten them to the applicable torque value.

    Task completed _____

27. Install the wheel assemblies.

    Task completed _____

28. Remove the safety stands from under the frame and axle.

    Task completed _____

29. Lower the truck back onto the floor.

    Task completed _____

30. If the radius rods have been loosened, or the equalizer bracket has been removed, check the rear alignment.

    Task completed _____

31. If necessary, adjust the axle alignment.

    Task completed _____

STUDENT SELF-EVALUATION

Check	Level	Competency	Comments
	4	Mastered task	
	3	Competent but need further help	
	2	Needed a lot of help	
	0	Did not understand the task	

INSTRUCTOR EVALUATION

Check	Level	Competency	Comments
	4	Mastered task	
	3	Competent but needs further help	
	2	Requires more training	
	0	Unable to perform task	

JOB SHEET 26.2

Name _____ Station _____ Date _____

Check a Height Control Valve.

Performance Objective(s): Check the operation of a height control valve on a truck air suspension system and test the components for air leaks.

ASE Education Foundation Correlation

This job sheet addresses the following ASE Education Foundation task(s):

IV. Suspension and Steering Systems
  A. General
   1. Research vehicle service information, including fluid type, vehicle service history, service precautions, and technical service bulletins. (IMMR, TST, MTST) P-1.
  E. Suspension Systems
   1. Inspect shock absorbers, bushings, brackets, and mounts; determine needed action. (IMMR) P-1.
   1. Inspect, service, repair, and/or replace shock absorbers, bushings, brackets, and mounts. (TST, MTST) P-1.
   3. Inspect axle and axle aligning devices such as: radius rods, track bars, stabilizer bars, and torque arms; inspect related bushings, mounts, and shims. (IMMR) P-1.
   3. Inspect, repair, and/or replace axle and axle aligning devices such as: radius rods, track bars, stabilizer bars, and torque arms; inspect related bushings, mounts, shims and attaching hardware; determine needed action. (TST, MTST) P-1.
   5. Inspect, repair, and/or replace air springs, mounting plates, springs, suspension arms, and bushings; replace as needed. (TST, MTST) P-1.
   6. Inspect air springs, mounting plates, springs, suspension arms, and bushings. (IMMR) P-1.
   5. Inspect and test air suspension pressure regulator and height control valves, lines, hoses, dump valves, and fittings; check and record ride height. (IMMR) P-1.
   6. Inspect, test, repair, and/or replace air suspension pressure regulator and height control valves, lines, hoses, dump valves, and fittings; check and record ride height. (TST, MTST) P-1.
   8. Measure, record and adjust ride height; determine needed action. (TST, MTST) P-1.

Tools and Materials: A truck or trailer equipped with an air suspension and height control valves, standard shop hand and power tools, and a soap-and-water solution

Protective Clothing: Standard shop apparel including coveralls or shop coat, safety glasses, and safety footwear

PROCEDURE
Ensure that the shop LOTO is observed.

There is always a reaction delay to position change in a height control valve. This may be as little as 2 seconds and as long as 15 seconds. Check the OEM specifications.

1. Park the vehicle on a level surface, ensuring that there is no suspension tension, and block the wheels.

   Task completed _____

2. Locate the height control valve to be tested. Remove the bolt that attaches the height control valve lever to the control rod.

   Task completed _____

3. Ensure that system air pressure is at least 100 psi (6.89 bar).

   Task completed _____

4. Raise the height control valve lever 45 degrees above horizontal.

   Task completed _____

5. Check for shock absorber extension to occur within 15 seconds depending on the reaction delay on the specific height control valve.

   Task completed _____

6. Next lower the height control valve lever 45 degrees below horizontal. This should cause air to bleed down within 15 seconds.

   Task completed _____

7. Release air until the air spring height drops to approximately 10 inches (254 mm).

   Task completed _____

8. Now raise the height control valve lever to 45 degrees above horizontal again and hold until the air spring height is approximately 12.5 inches (318 mm).

   Task completed _____

9. Release the valve lever and allow it to center.

   Task completed _____

10. Check the valve body, all hose connections, and air spring assemblies for leaks with soap-and-water solution.

    Task completed _____

11. Recheck the air spring height 15 minutes after performing step 8, noting whether the air spring height changed.

    Task completed _____

12. Now reinstall the fastener that attaches the height control lever to the height control rod.

    Task completed _____

13. Torque the nut to the OEM specification.

    Task completed _____

STUDENT SELF-EVALUATION

Check	Level	Competency	Comments
	4	Mastered task	
	3	Competent but need further help	
	2	Needed a lot of help	
	0	Did not understand the task	

INSTRUCTOR EVALUATION

Check	Level	Competency	Comments
	4	Mastered task	
	3	Competent but needs further help	
	2	Requires more training	
	0	Unable to perform task	

JOB SHEET 26.3

Name _____ Station _____ Date _____

Replace an Air Spring.

Performance Objective(s): Become familiar with the procedure required to replace an air spring assembly on a heavy-duty air suspension system.

ASE Education Foundation Correlation

This job sheet addresses the following ASE Education Foundation task(s):

IV. Suspension and Steering Systems
   A. General
   1. Research vehicle service information, including fluid type, vehicle service history, service precautions, and technical service bulletins. (IMMR, TST, MTST) P-1.

   E. Suspension Systems
   1. Inspect shock absorbers, bushings, brackets, and mounts; determine needed action. (IMMR) P-1.
   1. Inspect, service, repair, and/or replace shock absorbers, bushings, brackets, and mounts. (TST, MTST) P-1.
   3. Inspect axle and axle aligning devices such as: radius rods, track bars, stabilizer bars, and torque arms; inspect related bushings, mounts, and shims. (IMMR) P-1.
   3. Inspect, repair, and/or replace axle and axle aligning devices such as: radius rods, track bars, stabilizer bars, and torque arms; inspect related bushings, mounts, shims and attaching hardware; determine needed action. (TST, MTST) P-1.
   5. Inspect, repair, and/or replace air springs, mounting plates, springs, suspension arms, and bushings; replace as needed. (TST, MTST) P-1.
   6. Inspect air springs, mounting plates, springs, suspension arms, and bushings. (IMMR) P-1.
   5. Inspect and test air suspension pressure regulator and height control valves, lines, hoses, dump valves, and fittings; check and record ride height. (IMMR) P-1.
   6. Inspect, test, repair, and/or replace air suspension pressure regulator and height control valves, lines, hoses, dump valves, and fittings; check and record ride height. (TST, MTST) P-1.
   8. Measure, record and adjust ride height; determine needed action. (TST, MTST) P-1.

Tools and Materials: A truck with an air suspension, scissor jack, frame stands, and shop hand and power tools

Protective Clothing: Standard shop apparel including coveralls or shop coat, safety glasses, and safety footwear

PROCEDURE
Ensure that the shop LOTO is observed.

1. Check the load on the vehicle suspension. Exhaust the air pressure from the air spring.

   Task completed _____

2. Raise the truck frame to remove the load from the suspension. This may be possible by raising the air spring height control valve or using a scissor jack. Block the frame with safety stands.

   Task completed _____

3. Remove the locknuts and washers that connect the air spring to the air spring upper mounting bracket (reference Figure 26–3).

   Task completed _____

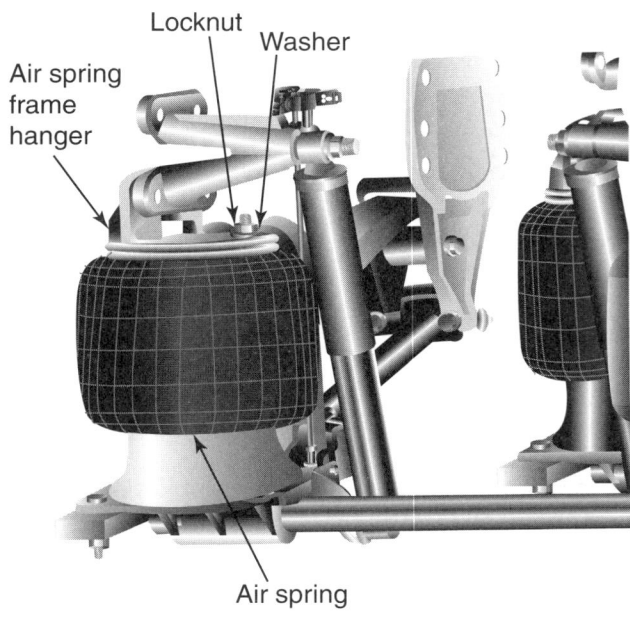

FIGURE 26–3 Air spring replacement

4. Remove the air lines connected to the air spring.

    Task completed _____

5. Remove the brass air fittings from the air spring.

    Task completed _____

6. Remove the locknuts and washers that connect the air spring pedestal to the frame pad.

    Task completed _____

7. Remove the air spring and pedestal assembly.

    Task completed _____

8. Install the replacement air spring to the upper mounting bracket by inserting the studs on the air spring into the appropriate holes on the hanger.

    Task completed _____

9. Install the air spring to the lower mounting bracket by inserting bolts into the holes of the pedestal assembly.

    Task completed _____

10. Install the locknuts and washers to secure the air spring to the bracket.

    Task completed _____

11. Tighten and torque the locknuts.

    Task completed _____

12. Install the locknuts and washers to secure the air spring to the frame hanger.

    Task completed _____

13. Tighten and torque the locknuts.

    Task completed _____

14. Install the brass air fitting to the air spring using a sealant.

    Task completed _____

15. Install the air lines to the brass fitting.

    Task completed _____

16. Supply air pressure to the air spring.

    Task completed _____

17. Adjust the height of the truck or trailer to specification using the height control valve.

    Task completed _____

STUDENT SELF-EVALUATION

Check	Level	Competency	Comments
	4	Mastered task	
	3	Competent but need further help	
	2	Needed a lot of help	
	0	Did not understand the task	

INSTRUCTOR EVALUATION

Check	Level	Competency	Comments
	4	Mastered task	
	3	Competent but needs further help	
	2	Requires more training	
	0	Unable to perform task	

ONLINE TASKS

Use a search engine and collect a list of bookmarks for truck and trailer suspension manufacturers. Here is some help to get you started:

1. Hendrickson
2. Neway Anchorlock
3. Reyco
4. Granning
5. Holland
6. Simard

STUDY TIPS

Identify 5 key points in Chapter 26. Try to be as brief as possible.

Key point 1 _____

Key point 2 _____

Key point 3 _____

Key point 4 _____

Key point 5 _____

27 Wheels and Tires

Objectives

After reading this chapter, you should be able to:

- Identify the wheel configurations used on heavy-duty trucks.
- Explain the difference between standard and wide-base wheel systems and stud- and hub-piloted mountings.
- Identify the common types of tire-to-rim hardware and describe their functions.
- Explain the importance of proper matching and assembly of tire and rim hardware.
- Outline the safety procedure for handling and servicing wheels and tires.
- Identify the different means of balancing tire and wheel assemblies.
- Describe brake drum mounting configurations.
- Perform wheel runout checks and adjustments.
- Properly match tires in dual and tandem mountings.
- List the major components of both grease- and oil-lubricated wheel hubs.
- Perform bearing and seal service on grease-lubricated front and rear wheel hubs.
- Perform bearing and seal service on oil-lubricated front and rear wheel hubs.
- Understand the concept behind axle-end dynamic balance and the consequences of an out-of-balance condition.
- Perform front and rear bearing adjustment.
- Describe TMC wheel end procedure.
- Outline the procedure for installing preset bearing wheels.

PRACTICE QUESTIONS

1. Technician A says that cast spoke wheels tend to produce more wheel alignment problems than disc wheels. Technician B says that DOT data shows that disc wheels produce more wheel-off incidents than cast spoke. Who is correct?
   a. Technician A only
   b. Technician B only
   c. both A and B
   d. neither A nor B

2. Technician A says that some disc wheels are lug- and stud mounted. Technician B says that stud pilot mount wheels are less common today due to the greater mounting safety of hub pilot mounts. Who is correct?
   a. Technician A only
   b. Technician B only
   c. both A and B
   d. neither A nor B

3. Technician A says that all cast spoke wheels have left-hand thread studs. Technician B says that stud-piloted mounts thread in a direction opposite to wheel rotation. Who is correct?
   a. Technician A only
   b. Technician B only
   c. both A and B
   d. neither A nor B

4. From the following images, identify the dual stud piloted fastener.
   a. A
   b. B

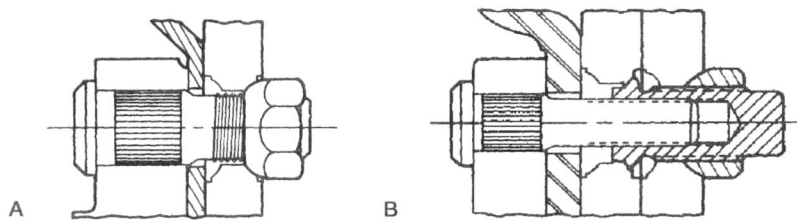

5. Technician A says that steel disc wheels can handle higher payloads than aluminum disc. Technician B says that because cast spoke wheels can handle more abuse than disc wheels, they are often used on rough terrain vocational trucks. Who is correct?
   a. Technician A only
   b. Technician B only
   c. both A and B
   d. neither A nor B

6. Radial and bias ply tires differ in their:
   a. dynamic footprint
   b. tread profile
   c. handling characteristics
   d. all of the above

7. What is the greatest allowable diameter variation permitted in a set of dual tires?
   a. 1/16 inch (1.6 mm)
   b. 1/8 inch (3 mm)
   c. 1/4 inch (6 mm)
   d. 1/2 inch (12.5 mm)

8. When a wheel bearing is described as being "wet," it is lubricated by:
   a. water
   b. oil
   c. grease
   d. silicone

9. What type of seal is shown in Figure 27–1?
   a. lip seal
   b. unitized seal
   c. barrier seal
   d. guardian seal

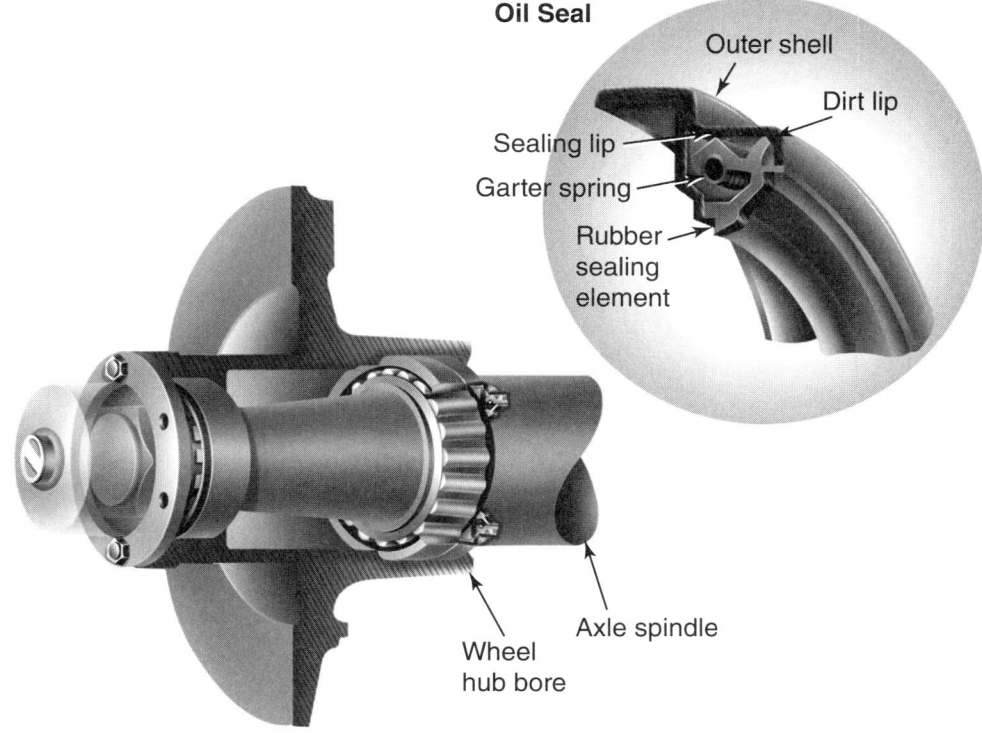

FIGURE 27–1

10. Technician A uses a Tempilstik crayon and an oxyacetylene torch to remove bearing cups from an aluminum wheel hub. Technician B uses a thermostat-controlled oven to remove races from aluminum hubs. Who is correct?
    a. Technician A only
    b. Technician B only
    c. both A and B
    d. neither A nor B

JOB SHEET 27.1

Name _____ Station _____ Date _____

Install a Set of Dual, Cast Spoke Wheels.

Performance Objective(s): Mount a pair of tire and rim assemblies to a cast spoke wheel assembly, and set runout and torque to specification.

ASE Education Foundation Correlation

This job sheet addresses the following ASE Education Foundation task(s):

IV. Suspension and Steering Systems
   A. General
   1. Research vehicle service information, including fluid type, vehicle service history, service precautions, and technical service bulletins. (IMMR, TST, MTST) P-1.

   G. Wheels and Tires
   3. Check wheel mounting hardware; check wheel condition; remove and install wheel/tire assemblies (steering and drive axle); torque fasteners to manufacturer's specification using torque wrench. (IMMR, TST, MTST) P-1.

Tools and Materials: A truck equipped with cast spoke wheels and standard shop hand and power tools

Protective Clothing: Standard shop apparel including coveralls or shop coat, safety glasses, and safety footwear

PROCEDURE
Ensure that the shop LOTO is observed.

This sequence assumes that the cast spoke wheel hub to be worked on is already raised and on stands and that the vehicle's wheels are blocked.

1. Ensure that the external spoke flats are clean and rust-free by using a wire brush. Slide the inner rear tire and rim assembly over the cast spoke wheel and back into position against the tapered mounting surface.

   Task completed _____

2. Position the rim on the wheel so that the valve stem faces out and is centered between the two spokes.

   Task completed _____

3. Slide the spacer ring over the wheel. This may have to be tapped into position. It should contact the inside tire and rim assembly.

   Task completed _____

4. Check the spacer ring for concentricity by rotating it around the cast spoke wheel. Replace it if there are any dents or visible runout irregularities.

   Task completed _____

5. Lift and slide the outside rear tire and rim assembly onto the wheel.

   Task completed _____

6. Align the rim so that the valve stem faces inboard and is located in the same relative position as the inner valve stem.

   Task completed _____

7. Assemble all rim clamps and nuts. Turn the nuts on their studs until each nut is flush with its clamp.

   Task completed _____

8. Reference Figure 27–2 and turn the top nut (1) until it is snug with the clamp.

Task completed _____

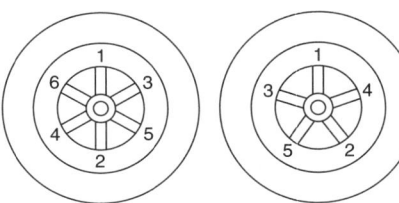

FIGURE 27–2 Tightening and torquing sequence for cast spoke wheels.

9. Rotate the wheel and rim until nut 2 is at the top position and snug the nut.

Task completed _____

10. Rotate the wheel and rim until nuts 3, 4, 5, and 6 are at the top, respectively, and snug these nuts.

Task completed _____

11. Now check the runout. Use a tire hammer face down on the floor so that the cross peen is located ⅛ inch from the tire, and then rotate the wheel assembly. Alternately loosen and tighten the tire nuts to remove excessive runout. Hitting the rims with a tire hammer can help this process, but care should be taken to avoid damaging the rims.

Task completed _____

12. When the runout is acceptable, torque the nuts in sequence to the manufacturer's recommended torque.

Task completed _____

CAUTION! Using any kind of pneumatic wrench when final-torquing wheel fasteners has become unacceptable due to the number of wheel-off incidents occurring in the United States and Canada. Torque values for cast spoke rims are surprisingly low, typically 175 to 250 lb-ft. (235 to 340 N·m), and exceeding these values can damage studs and tire rims.

STUDENT SELF-EVALUATION

Check	Level	Competency	Comments
	4	Mastered task	
	3	Competent but need further help	
	2	Needed a lot of help	
	0	Did not understand the task	

INSTRUCTOR EVALUATION

Check	Level	Competency	Comments
	4	Mastered task	
	3	Competent but needs further help	
	2	Requires more training	
	0	Unable to perform task	

JOB SHEET 27.2

Name _____ Station _____ Date _____

Remove a Front Disc Wheel Hub Assembly.

Performance Objective(s): Become familiar with the process for removing a front disc wheel hub assembly from a truck.

ASE Education Foundation Correlation

This job sheet addresses the following ASE Education Foundation task(s):

IV. **Suspension and Steering Systems**
    A. **General**
    1. Research vehicle service information, including fluid type, vehicle service history, service precautions, and technical service bulletins. (IMMR, TST, MTST) P-1.

    G. **Wheels and Tires**
    3. Check wheel mounting hardware; check wheel condition; remove and install wheel/tire assemblies (steering and drive axle); torque fasteners to manufacturer's specification using torque wrench. (IMMR, TST, MTST) P-1.

Tools and Materials: A truck equipped with steel or aluminum disc wheels, jacking and hoisting equipment, stands, and standard shop hand and power tools

Protective Clothing: Standard shop apparel including coveralls or shop coat, safety glasses, and safety footwear

PROCEDURE
Ensure that the shop LOTO is observed.

1. Block enough wheels on the truck to prevent vehicle movement and set the parking brakes.

    Task completed _____

2. Jack the front of the truck until the tires clear the ground.

    Task completed _____

3. Place safety stands under the front axle and lower the truck so that its weight is supported by the safety stands.

    Task completed _____

4. Fully back off the slack adjuster to produce maximum shoe-to-drum clearance. Check how this is accomplished when auto-slacks are used.

    Task completed _____

5. Remove the wheel fasteners. Ensure that the nuts are rotated in the correct direction to remove them. Pull the disc wheel assembly.

    Task completed _____

6. Remove the brake drum.

    Task completed _____

7. Remove the hub cap, hub cap gasket, jam or locknut, lock washer, and lock ring.

    Task completed _____

8. Back off the wheel bearing adjusting nut about two turns or enough to allow the weight of the hub to be lifted from the wheel bearings.

Task completed _____

9. Lift the hub until all weight is removed from the wheel bearing.

Task completed _____

10. Remove the adjusting nut.

Task completed _____

11. Move the hub about ½ inch (12.5 mm) to jar loose the outer wheel bearing.

Task completed _____

12. Carefully lift the outer bearing off the axle. Do not allow it to drop.

Task completed _____

13. Remove the hub from the axle spindle.

Task completed _____

14. Remove the inner wheel bearing and bearing spacer (if used) from the axle.

Task completed _____

15. Remove the seal from the axle if it has not already been removed.

Not applicable _____

Task completed _____

STUDENT SELF-EVALUATION

Check	Level	Competency	Comments
	4	Mastered task	
	3	Competent but need further help	
	2	Needed a lot of help	
	0	Did not understand the task	

INSTRUCTOR EVALUATION

Check	Level	Competency	Comments
	4	Mastered task	
	3	Competent but needs further help	
	2	Requires more training	
	0	Unable to perform task	

JOB SHEET 27.3

Name _____ Station _____ Date _____

Reassemble a Disc Wheel Hub Assembly.

Performance Objective(s): Become familiar with the procedure for mounting a hub assembly on a disc wheel assembly.

ASE Education Foundation Correlation

This job sheet addresses the following ASE Education Foundation task(s):

IV. Suspension and Steering Systems
  A. General
  1. Research vehicle service information, including fluid type, vehicle service history, service precautions, and technical service bulletins. (IMMR, TST, MTST) P-1.

  G. Wheels and Tires
  3. Check wheel mounting hardware; check wheel condition; remove and install wheel/tire assemblies (steering and drive axle); torque fasteners to manufacturer's specification using torque wrench. (IMMR, TST, MTST) P-1.

Tools and Materials: A truck equipped with steel or aluminum disc wheels, jacking and hoisting equipment, stands, and standard shop hand and power tools

Protective Clothing: Standard shop apparel including coveralls or shop coat, safety glasses, and safety footwear

PROCEDURE
Ensure that the shop LOTO is observed.

This procedure assumes that the disc wheel assembly has been disassembled as per the procedure in Job Sheet Number 27.2.

1. If reassembling an aluminum disc wheel, the assembly procedure may require that it be preheated. This should be performed in a thermostat-controlled heating oven.

   Task completed _____

2. Use a mounting dolly or sleeve to press the bearing race or cup into the hub. Take extra care when working with aluminum disc wheels to avoid damage.

   Task completed _____

3. Ensure that the bearing race fully seats into the bearing bore.

   Task completed _____

4. Lubricate the wheel bearings with gear oil. With dry lube wheels, this may require packing the hub cavity between the two bearing cups with wheel bearing grease to the level of the cup's smallest diameter. Hand or pressure packing of greased wheel bearings is required.

   Task completed _____

5. Insert the inner bearing cone into the hub.

   Task completed _____

6. Prelubricate the seal by wiping it with the lubricant.

   Task completed _____

7. Place the prelubed seal in the hub with the lip facing the bearing cone.

   Task completed _____

8. Drive the seal into its recess with the correct size seal driver.

   Task completed _____

9. Position the spacer (if used) on the spindle. Align the hole and pin.

   Task completed _____

10. Apply a light film of lubricant to the spindle.

    Task completed _____

11. Use a wheel dolly to center the wheel/hub assembly on the spindle.

    Task completed _____

12. Push the wheel/hub onto the spindle far enough so that the seal rides over its contact surface on the bearing spacer or spindle.

    Task completed _____

13. Install the outer bearing cone, washer, and adjusting nut in reverse order of removal.

    Task completed _____

14. Adjust the bearing according to the TMC procedure.

    Task completed _____

15. Secure the locknut and locking device.

    Task completed _____

16. Check the lubricant in the hub cap.

    Task completed _____

17. Position the new gasket with sealant on the hub cap.

    Task completed _____

18. Install the hub cap. In a wet lube system, fill to the correct lubricant level.

    Task completed _____

STUDENT SELF-EVALUATION

Check	Level	Competency	Comments
	4	Mastered task	
	3	Competent but need further help	
	2	Needed a lot of help	
	0	Did not understand the task	

INSTRUCTOR EVALUATION

Check	Level	Competency	Comments
	4	Mastered task	
	3	Competent but needs further help	
	2	Requires more training	
	0	Unable to perform task	

JOB SHEET 27.4

Name _____ Station _____ Date _____

Learn the TMC Recommended Wheel-End Adjustment Procedure.

Performance Objective(s): Learn the TMC Recommended Wheel-End Adjustment Procedure exactly as it appears in the Recommended Practice. This must be used on all wheel ends not classified as unitized.

ASE Education Foundation Correlation
This job sheet addresses the following ASE Education Foundation task(s):

III. Brakes
  J. Wheel Bearings
  1. Clean, inspect, lubricate, and/or replace wheel bearings and races/cups; replace seals and wear rings; inspect spindle/tube; inspect and replace retaining hardware; adjust wheel bearings; check hub assembly fluid level and condition; verify end play with dial indicator method. (IMMR, TST, MTST) P-1.
  2. Identify, inspect, and/or replace unitized/preset hub bearing assemblies. (IMMR, TST, MTST) P-2.

IV. Suspension and Steering Systems
  A. General
  1. Research vehicle service information, including fluid type, vehicle service history, service precautions, and technical service bulletins. (IMMR, TST, MTST) P-1.

  G. Wheels and Tires
  2. Identify wheel/tire vibration, shimmy, pounding, and hop (tramp) problems. (IMMR) P-2.
  2. Identify wheel/tire vibration, shimmy, pounding, and hop (tramp) problems; determine needed action. (TST) P-2.
  2. Diagnose wheel/tire vibration, shimmy, pounding, and hop (tramp) problems; determine needed action. (MTST) P-2.

Tools and Materials: A truck or trailer with the wheel end disassembled, a torque wrench, and axle end sockets

Protective Clothing: Standard shop apparel including coveralls or shop coat, safety glasses, and safety footwear

PROCEDURE
Ensure that the shop LOTO is observed.

This seven-step bearing adjustment procedure was developed by the TMC's Wheel End task force with input from all the OEM bearing manufacturers. The Standard values used in TMC RP 618 have not been converted into metric equivalents.

1. Bearing Lubrication. Lubricate the wheel bearing with clean lubricant of the same type used in the axle sump or hub assembly.

   Task completed _____

2. Initial Adjusting Nut Torque. Tighten the adjusting nut to a torque of 200 lb-ft. while rotating the wheel.

   Task completed _____

3. Initial Back-Off. Back off the adjusting nut one full turn.

   Task completed _____

4. Final Adjusting Nut Torque. Tighten the adjusting nut to a final torque of 50 lb-ft. while rotating the wheel.

   Task completed _____

5. Final Back-Off.

Axle Type	Threads per Inch	Final Back-Off
Steer single nut	12	1/6 turn—cotter pin lock
Steer single nut	18	1/4 turn—cotter pin lock
Steer double nut	14	1/2 turn
Steer double nut	18	1/2 turn
Drive	12	1/4 turn
Drive	16	1/4 turn
Trailer	12	1/4 turn
Trailer	16	1/4 turn

Task completed _____

6. Jam Nut Torque.

Axle Type	Nut Size	Torque Specification
Steer double nut	Less than 2 5/8 inch	200–300 lb-ft.
Steer double nut	2 5/8 inch and over	300–400 lb-ft.
Drive	Dowel-type washer	300–400 lb-ft.
Drive	Tang-type washer	200–275 lb-ft.
Trailer	Less than 2 5/8 inch	200–300 lb-ft.
Trailer	2 5/8 inch and over	300–400 lb-ft.

Task completed _____

7. Measure Endplay. Attach a magnetic-base dial indicator to the hub or brake drum. Adjust the dial indicator so that its plunger is contacting the spindle end with its line of action parallel to the axis of the spindle. Grasp the wheel or hub assembly by hand in the 3 o'clock and 9 o'clock positions. Alternately thrust and pull the wheel while oscillating the wheel through 45 degrees of travel. Stop oscillating the hub so that the dial indicator tip is in the same position as it was before the oscillation began. Read the bearing endplay as total indicator movement.

Acceptable endplay is 0.001–0.005 inch. If there is either zero endplay or more than this specification, repeat the wheel-end adjustment procedure.

Task completed _____

STUDENT SELF-EVALUATION

Check	Level	Competency	Comments
	4	Mastered task	
	3	Competent but need further help	
	2	Needed a lot of help	
	0	Did not understand the task	

INSTRUCTOR EVALUATION

Check	Level	Competency	Comments
	4	Mastered task	
	3	Competent but needs further help	
	2	Requires more training	
	0	Unable to perform task	

ONLINE TASKS

Use a search engine and check out the following OEM Web sites. Log them into your bookmark folders.

1. Accurride
2. Arvin-Meritor
3. Budd
4. Chicago Rawhide
5. Stemco
6. Michelin
7. Goodyear
8. Bridgestone
9. Bandag
10. Timken

STUDY TIPS

Identify 5 key points in Chapter 27. Try to be as brief as possible.

Key point 1 _____

Key point 2 _____

Key point 3 _____

Key point 4 _____

Key point 5 _____

28 Truck Brake Systems

Objectives

After reading this chapter, you should be able to:
- Identify the components of a truck air brake system.
- Explain the operation of a dual-circuit air brake system.
- Understand what is meant by pneumatic and torque imbalance.
- Describe the role played by the Federal Motor Vehicle Safety Standard No. 121 (FMVSS No. 121) on present-day air brake systems.
- Identify the major components and systems of an air compressor.
- Outline the operating principles of the valves and controls used in an air brake system.
- Explain the operation of an air brake chamber.
- Outline the functions of the hold-off and service circuits in truck and trailer brake systems.
- Describe the operation of S-cam and wedge-actuated drum brakes.
- Describe the operating principles of slack adjusters.
- List the components and describe the operating principles of an air disc brake system.
- Outline the factors that determine brake balance in truck rigs.

PRACTICE QUESTIONS

1. Technician A says that disc brakes use a steel rotor rotated at wheel speed that is squeezed on either side by friction pads on a stationary caliper when actuated. Technician B says that the clamping force exerted by the caliper in a pneumatic disc brake system is delivered by an air brake chamber. Who is correct?
   a. Technician A only
   b. Technician B only
   c. both A and B
   d. neither A nor B

2. Of the following, which is not an advantage of using compressed air over hydraulic fluid pressure?
   a. Air brakes remain effective even when substantial system leakage occurs.
   b. Compressed air is an effective control medium.
   c. Air is virtually limitless in supply.
   d. Air brake systems provide reduced service application timing lag over hydraulic brakes.

3. Which of the following would be considered part of a truck service brake system?
   a. foundation brakes
   b. treadle valve
   c. relay valves
   d. spring brake chambers

4. Which type of air compressor would be most common on today's heavy-duty trucks?
   a. reciprocating piston, 2-cylinder
   b. rotary vane
   c. swashplate, single cylinder
   d. reciprocating piston, 4-cylinder

5. The spring brake system provides which of the following functions?
   a. service braking
   b. emergency braking
   c. parking braking
   d. all of the above
6. Which of the following service application valves can be referred to as a trolley valve?
   a. trailer hand valve
   b. trailer supply valve
   c. system park valve
   d. treadle valve
7. Which of the following air brake system components could affect the pneumatic timing of service applications?
   a. valve crack pressure
   b. line sizes
   c. fitting restrictions
   d. all of the above
8. Technician A says that the compressor crankcase contains the crankshaft, rod, and main bearings. Technician B says that the compressor cylinder head assembly contains cam-actuated discharge and inlet valves. Who is correct?
   a. Technician A only
   b. Technician B only
   c. both A and B
   d. neither A nor B
9. Which of the following is the most common means of supplying oil to a flange-mounted air compressor?
   a. through the rear end cover
   b. through the cylinder head
   c. through the base of the crankcase
   d. through an oil supply tube directly to the crankshaft
10. Which of the following best describes the primary function of a tractor protection valve?
    a. protects the tractor air supply should trailer breakaway occur
    b. senses when the tractor is running without a trailer
    c. limits service application pressure to the tractor during normal service braking
    d. speeds up the exhausting of service application air from the trailer air chambers

JOB SHEET 28.1

Name _____ Station _____ Date _____

Air Brake System Component Identification.

Performance Objective(s): Familiarization with the critical valves and components on a truck air brake system

ASE Education Foundation Correlation

This job sheet addresses the following ASE Education Foundation task(s):

III. Brakes
   A. General
   1. Research vehicle service information, including fluid type, vehicle service history, service precautions, and technical service bulletins. (IMMR, TST, MTST) P-1.
   2. Identify brake system components and configurations (including air and hydraulic systems, parking brake, power assist, and vehicle dynamic brake systems). (IMMR, TST, MTST) P-1.

   C. Air Brakes: Mechanical/Foundation Brake System
   2. Identify slack adjuster type; inspect slack adjusters; determine needed action. (IMMR) P-1.
   2. Identify slack adjuster type; inspect slack adjusters; perform needed action. (TST, MTST) P-1.

Tools and Materials: A straight truck or tractor equipped with air brakes, a creeper, and a trouble light, and paper, colored markers, and a pencil

Protective Clothing: Standard shop apparel including coveralls or shop coat, safety glasses, and safety footwear

PROCEDURE

Use Figure 28–1 to help you identify the valves in the following table, filling in the blank fields where possible. The objective is to identify some critical valves on a truck air brake system and verify their operation.

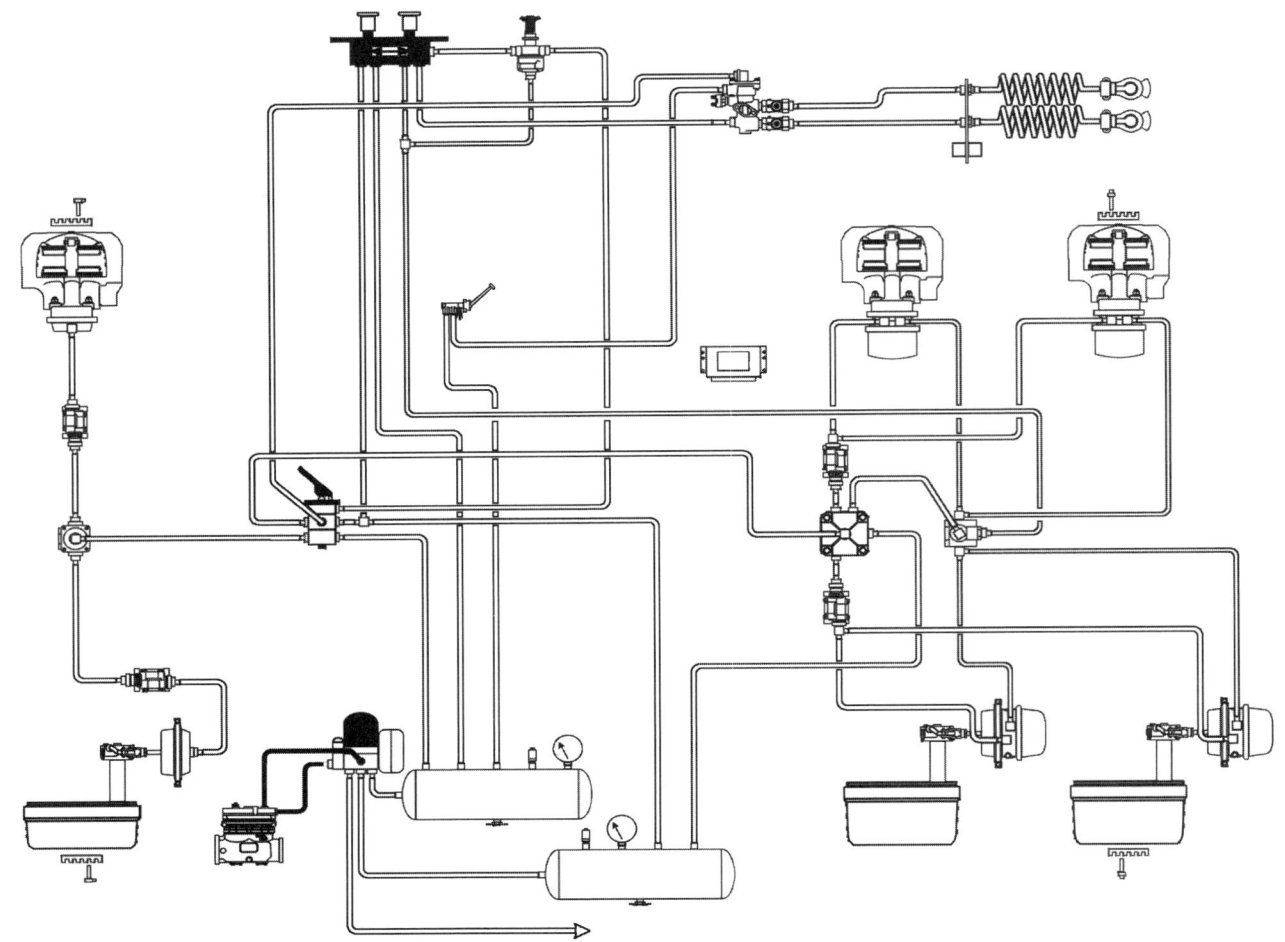

FIGURE 28–1 Typical brake circuit and components.

Valve/Component	Type	Manufacturer	Operation
Foot valve			
Dash control valves			
Governor			
Front axle service			
Rear axle service			
Park valve system			
Tractor/protection			
Trailer spike			
Governor			

Task completed _____

STUDENT SELF-EVALUATION

Check	Level	Competency	Comments
	4	Mastered task	
	3	Competent but need further help	
	2	Needed a lot of help	
	0	Did not understand the task	

INSTRUCTOR EVALUATION

Check	Level	Competency	Comments
	4	Mastered task	
	3	Competent but needs further help	
	2	Requires more training	
	0	Unable to perform task	

JOB SHEET 28.2

Name _____ Station _____ Date _____

Identify Foundation Brake Components.

Performance Objective(s): Disassemble a truck S-cam foundation brake on one wheel and identify the components.

ASE Education Foundation Correlation

This job sheet addresses the following ASE Education Foundation task(s):

III. Brakes
A. General
1. Research vehicle service information, including fluid type, vehicle service history, service precautions, and technical service bulletins. (IMMR, TST, MTST) P-1.
2. Identify brake system components and configurations (including air and hydraulic systems, parking brake, power assist, and vehicle dynamic brake systems). (IMMR, TST, MTST) P-1.

B. Air Brakes: Air Supply and Service Systems
3. Demonstrate knowledge and understanding of air supply and service system components and operations. (TST, MTST) P-1.

C. Air Brakes: Mechanical/Foundation Brake System
2. Identify slack adjuster type; inspect slack adjusters; determine needed action. (IMMR) P-1.
2. Identify slack adjuster type; inspect slack adjusters; perform needed action. (TST, MTST) P-1.
3. Check camshafts (S-cams), tubes, rollers, bushings, seals, spacers, retainers, brake spiders, shields, anchor pins, and springs; determine needed action. (IMMR) P-1.
3. Check camshafts (S-cams), tubes, rollers, bushings, seals, spacers, retainers, brake spiders, shields, anchor pins, and springs; perform needed action. (TST, MTST) P-1.
6. Remove brake drum; clean and inspect brake drum and mounting surface; measure brake drum diameter; measure brake lining thickness; inspect brake lining condition; determine needed action. (IMMR, TST, MTST) P-1.
7. Identify concerns related to the mechanical/foundation brake system including poor stopping, brake noise, premature wear, pulling, grabbing, or dragging; determine needed action. (TST) P-1.
7. Diagnose concerns related to the mechanical/foundation brake system including poor stopping, brake noise, premature wear, pulling, grabbing, or dragging; determine needed action. (MTST) P-1.

D. Air Brakes: Parking Brake System
1. Inspect and check parking (spring) brake chamber for leaks; determine needed action. (IMMR) P-1.
1. Inspect, test, and/or replace parking (spring) brake chamber. (TST, MTST) P-1.
2. Inspect and test parking (spring) brake check valves, lines, hoses, and fittings; determine needed action. (IMMR) P-1.
2. Inspect, test, and/or replace parking (spring) brake check valves, lines, hoses, and fittings. (TST, MTST) P-1.
3. Inspect and test parking (spring) brake application and release valve; determine needed action. (IMMR) P-1.
3. Inspect, test, and/or replace parking (spring) brake application and release valve. (TST, MTST) P-1.
5. Identify and test anti-compounding brake function; determine needed action. (TST, MTST) P-2.

F. Hydraulic Brakes: Mechanical/Foundation Brake System
1. Inspect rotor and mounting surface; measure rotor thickness, thickness variation, and lateral runout; determine needed action. (IMMR, TST) P-1.
1. Clean and inspect rotor and mounting surface; measure rotor thickness, thickness variation, and lateral runout; determine necessary action. (MTST) P-1.
2. Inspect and clean disc brake caliper assemblies; inspect and measure disc brake pads; inspect mounting hardware; determine needed action. (IMMR, TST, MTST) P-1.
3. Remove brake drum; clean and inspect brake drum and mounting surface; measure brake drum diameter; measure brake lining thickness; inspect brake lining condition; inspect wheel cylinders; determine needed action. (IMMR, TST, MTST) P-1.

H. Power Assist Systems
1. Check brake assist/booster system (vacuum or hydraulic) hoses and control valves; check fluid level and condition (if applicable). (IMMR, TST, MTST) P-1.

I. Vehicle Dynamic Brake Systems (Air and Hydraulic): Antilock Brake System (ABS), Automatic Traction Control (ATC) System, and Electronic Stability Control (ESC) System
1. Observe antilock brake system (ABS) warning light operation including trailer and dash mounted trailer ABS warning light.; determine needed action. (IMMR, TST, MTST) P-1.
2. Observe automatic traction control (ATC) and electronic stability control (ESC) warning light operation; determine needed action. (IMMR, TST, MTST) P-2.
8. Verify power line carrier (PLC) operation. (TST, MTST) P-3.

Tools and Materials: A truck or trailer equipped with S-cam drum brakes that can be disassembled, jacks and stands, a wheel dolly, shoe levers, and standard shop hand and power tools

Protective Clothing: Standard shop apparel including coveralls or shop coat, safety glasses, and safety footwear

PROCEDURE

1. Jack up the wheel to be disassembled and place it securely on stands.

    Task completed _____

2. Pull the wheel hub, making provision to catch the oil. If the wheel is on a drive axle, the axle shafts will have to be pulled.

    Task completed _____

3. Remove the outboard wheel bearing.

    Task completed _____

4. Place a wheel dolly under the wheel assembly and locate the forks so that the weight is evenly distributed.

    Task completed _____

5. Pull the wheel assembly away from the axle.

    Task completed _____

6. Observe the foundation brake assembly while it is assembled and make a note of anything that could cause problems.

    Task completed _____

7. Disassemble the foundation brake assembly, stripping the spider but leaving the S-cam assembly intact.

    Task completed _____

8. Fill in the blank fields in the following table:

Component	Type/Description	Condition/Wear/Play Spec
Brake shoes/friction faces		
S-cam		
S-cam bushings		
Retraction springs		
Cam rollers/anchor pins		
Axle spider		

Task completed _____

9. Reassemble the foundation brake assembly, replacing any defective components.

Task completed _____

10. Reassemble the wheel assembly using the TMC method of adjusting the wheel end.

Task completed _____

11. Ensure that the wheel end is properly lubricated.

Task completed _____

STUDENT SELF-EVALUATION

Check	Level	Competency	Comments
	4	Mastered task	
	3	Competent but need further help	
	2	Needed a lot of help	
	0	Did not understand the task	

INSTRUCTOR EVALUATION

Check	Level	Competency	Comments
	4	Mastered task	
	3	Competent but needs further help	
	2	Requires more training	
	0	Unable to perform task	

ONLINE TASKS

Using a search engine, identify the major manufacturers of truck brake systems and note what they have to say online about their products. Here are some clues:

1. Bendix
2. Haldex
3. Arvin Meritor
4. MGMbrake chambers

STUDY TIPS

Identify 5 key points in Chapter 28. Try to be as brief as possible.

Key point 1 _____

Key point 2 _____

Key point 3 _____

Key point 4 _____

Key point 5 _____

29 Hydraulic Brakes and Air-over-Hydraulic Brake Systems

Objectives

After reading this chapter, you should be able to:
- Describe the principles of operation of a hydraulic brake system.
- Identify the major components in a truck hydraulic brake system.
- Describe the operation of drum brakes in a hydraulic braking system.
- Describe the operation of wheel cylinders and calipers.
- Identify the different types of disc brake calipers.
- List the major components of a master cylinder.
- Identify the hydraulic valves and controls used in hydraulic brake systems.
- Explain the operation of a hydraulic power booster.
- List the major components of an air-over-hydraulic braking system.
- Outline some typical maintenance and service procedures performed on hydraulic and air-over-hydraulic brake systems.
- Bench bleed a master cylinder.
- Identify the methods used to bleed brakes.
- Describe the operation of a typical hydraulic ABS.

PRACTICE QUESTIONS

1. Which of the following components in a truck hydraulic brake system should have no direct contact with the brake fluid?
   a. master cylinder
   b. caliper piston
   c. metering valve
   d. brake pads

2. Which of the following components would more likely be found in a split drum-and-disc hydraulic brake system and not in a drum-only system?
   a. master cylinder
   b. hydraulic lines and hoses
   c. power booster control valves
   d. metering valve

3. Which component of a hydraulic brake system regulates pressure applied to the rear drum brakes?
   a. proportioning valve
   b. combination valve
   c. metering valve
   d. pressure differential valve

4. Which of the following describes the role of a hydraulic brake system load proportioning valve?
   a. distributes the truck load so that brake imbalance is minimized
   b. proportions hydraulic pressure to the front and rear brakes based on vehicle loads
   c. balances braking when truck load is unevenly distributed
   d. all of the above

5. Which of the following types of parking brakes would be most commonly used in an air-over-hydraulic system?
   a. canister-type spring brake
   b. ratchet type
   c. driveline drum
   d. hydraulic piston

6. When the term *hygroscopic* is used to describe a brake fluid, what does it mean?
   a. tending to absorb moisture from water
   b. tending to absorb moisture from air
   c. subject to high-temperature breakdown
   d. subject to coagulation

7. Which of the following is a disadvantage of a hydraulic brake system when compared to an air brake system?
   a. application timing
   b. release timing
   c. ability to function effectively when system leakage occurs
   d. brake feel

8. In a dual-circuit hydraulic brake system, which of the following valves switches the brake warning light in the event of severe fluid leakage in one of the circuits?
   a. metering valve
   b. pressure differential valve
   c. proportioning valve
   d. wheel cylinder

9. Which of the following hydraulic brake configurations would likely provide the highest braking efficiencies on a two-axle straight truck chassis?
   a. four-wheel disc
   b. four-wheel drum
   c. front disc, rear drum
   d. front drum, rear disc

10. Which of the following is a disadvantage of disc brakes over drum brakes?
    a. ability to automatically adjust
    b. application speed
    c. requirement of higher hydraulic pressures
    d. susceptibility to dirt accumulation

JOB SHEET 29.1

Name _____ Station _____ Date _____

Pressure Bleed a Hydraulic Brake System.

Performance Objective(s): Become familiar with the procedure required to pressure bleed a hydraulic brake circuit using a pressure bleeder.

ASE Education Foundation Correlation

This job sheet addresses the following ASE Education Foundation task(s):

III. Brakes
A. General
1. Research vehicle service information, including fluid type, vehicle service history, service precautions, and technical service bulletins. (IMMR, TST, MTST) P-1.
2. Identify brake system components and configurations (including air and hydraulic systems, parking brake, power assist, and vehicle dynamic brake systems). (IMMR, TST, MTST) P-1.
3. Identify brake performance problems caused by the mechanical/foundation brake system. (air and hydraulic) (IMMR, TST, MTST) P-1.
4. Use appropriate electronic service tool(s) and procedures to diagnose problems; check, record, and clear diagnostic codes; interpret digital multimeter (DMM) readings. (TST, MTST) P-1.

E. Hydraulic Brakes: Hydraulic System
1. Check master cylinder fluid level and condition; determine proper fluid type for application. (IMMR, TST, MTST) P-1.
2. Inspect hydraulic brake system components for leaks and damage. (IMMR) P-1.
2. Inspect hydraulic brake system for leaks and damage; test, repair, and/or replace hydraulic brake system components. (TST, MTST) P-1.
3. Check hydraulic brake system operation including pedal travel, pedal effort, and pedal feel. (IMMR) P-1.
3. Check hydraulic brake system operation including pedal travel, pedal effort, and pedal feel; determine needed action. (TST, MTST) P-1.
4. Identify poor stopping, premature wear, pulling, dragging, imbalance, or poor pedal feel caused by problems in the hydraulic system; determine needed action. (TST) P-2.
4. Diagnose poor stopping, premature wear, pulling, dragging, imbalance, or poor pedal feel caused by problems in the hydraulic system; determine needed action. (MTST) P-2.
5. Test master cylinder for internal/external leaks and damage; replace as needed. (TST, MTST) P-2.
6. Test metering (hold-off), load sensing/proportioning, proportioning, and combination valves; determine needed action. (TST, MTST) P-3.
7. Test brake pressure differential valve; test warning light circuit switch, bulbs/LEDs, wiring, and connectors; determine needed action. (TST, MTST) P-2.
8. Bleed and/or flush hydraulic brake system. (TST, MTST) P-2.

F. Hydraulic Brakes: Mechanical/Foundation Brake System
1. Inspect rotor and mounting surface; measure rotor thickness, thickness variation, and lateral runout; determine needed action. (IMMR, TST) P-1.
1. Clean and inspect rotor and mounting surface; measure rotor thickness, thickness variation, and lateral runout; determine necessary action. (MTST) P-1.
2. Inspect and clean disc brake caliper assemblies; inspect and measure disc brake pads; inspect mounting hardware; determine needed action. (IMMR) P-1.
2. Inspect and clean disc brake caliper assemblies; inspect and measure disc brake pads; inspect mounting hardware; perform needed action. (TST, MTST) P-1.
3. Remove brake drum; clean and inspect brake drum and mounting surface; measure brake drum diameter; measure brake lining thickness; inspect brake lining condition; inspect wheel cylinders; determine needed action. (IMMR, TST, MTST) P-1.

H. Power Assist Systems
1. Check brake assist/booster system (vacuum or hydraulic) hoses and control valves; check fluid level and condition (if applicable). (IMMR, TST, MTST) P-1.

I. Vehicle Dynamic Brake Systems (Air and Hydraulic): Antilock Brake System (ABS), Automatic Traction Control (ATC) System, and Electronic Stability Control (ESC) System
1. Observe antilock brake system (ABS) warning light operation including trailer and dash-mounted trailer ABS warning light. (IMMR) P-1.
1. Observe antilock brake system (ABS) warning light operation including trailer and dash-mounted trailer ABS warning light; determine needed action. (TST, MTST) P-1.
2. Observe automatic traction control (ATC) and electronic stability control (ESC) warning light operation. (IMMR) P-2.
2. Observe automatic traction control (ATC) and electronic stability control (ETC) warning light operation; determine needed action. (TST, MTST) P-2.
3. Identify stopping concerns related to the vehicle dynamic brake systems: ABS, ATC, and ESC; determine needed action. (TST, MTST) P-2.
5. Check and test operation of vehicle dynamic brake system (air and hydraulic) mechanical and electrical components; determine needed action. (TST, MTST) P-1.
6. Test vehicle/wheel speed sensors and circuits; adjust, repair, and/or replace as needed. (TST, MTST) P-1.
7. Bleed ABS hydraulic circuits. (TST, MTST) P-2.
8. Verify power line carrier (PLC) operation. (TST, MTST) P-3.

J. Wheel Bearings
1. Clean, inspect, lubricate, and/or replace wheel bearings and races/cups; replace seals and wear rings; inspect spindle/tube; inspect and replace retaining hardware; adjust wheel bearings; check hub assembly fluid level and condition; verify end play with dial indicator method. (IMMR, TST, MTST) P-1.
2. Identify, inspect, and/or replace unitized/preset hub bearing assemblies. (IMMR, TST, MTST) P-2.

Tools and Materials: A truck equipped with hydraulic brakes, a bleeder hose, brake fluid, a fluid receptacle, a pressure bleeder and adapter, and service literature

Protective Clothing: Standard shop apparel including coveralls or shop coat, safety glasses, and safety footwear

PROCEDURE
Ensure that the shop LOTO is observed.

This procedure assumes that a split disc/drum brake system is being bled.

1. Locate the master cylinder in the vehicle and obtain the pressure bleeder manufacturer's service instructions.

    Task completed _____

2. Remove the master cylinder cover and diaphragm, if fitted.

    Task completed _____

3. Fit the pressure bleeder adapter cover to the master cylinder using the correct adapter couplers. Ensure that there is sufficient brake fluid in the bleeder ball.

    Task completed _____

4. Raise the system pressure to the specified test pressure. The test pressure will vary depending on the manufacturer of the bleeder ball. Around 20 psi is typical.

    Task completed _____

5. Begin by bleeding the rear wheel cylinders first. Select a suitably sized box wrench and fit the bleeder hose to the bleeder fitting on the right rear wheel. Open the bleeder ball valve to pressurize the master cylinder and then crack the right rear bleeder fitting. With one end of the bleeder hose immersed in the fluid receptacle (a glass jar will do), crack the bleeder fitting and allow fluid to flow into it. When bubble-free fluid is exiting the hose, tighten the bleeder fitting.

    Task completed _____

6. Repeat step 5 on the left rear wheel.

    Task completed _____

7. Move to the front right disc brake caliper, locate the bleeder nipple, and fit the hose to it. To perform this procedure, the metering valve may have to be locked into an open position. Consult the OEM service literature for the procedure required to do this. Bleed down the front right wheel caliper.

    Task completed _____

8. Perform step 5 on the front left caliper.

    Task completed _____

9. This procedure began with bleeding out the circuit farthest from the master cylinder and finished with the portion closest. Pressure-boosted systems will require additional bleeding, so check OEM instructions to perform this. In specialty vocational trucks using right-hand steering, check the location of the master cylinder and sequence the bleeding of the circuit accordingly.

    Task completed _____

10. After completing the circuit bleeding, close the bleeder ball control valve and remove it from the master cylinder.

    Task completed _____

11. Check that disc brake pistons have returned to their normal positions and that the shoes are properly seated. Perform this by actuating the brake pedal a number of times.

    Task completed _____

12. Replace the diaphragm and check the fill level in the master cylinder before installing the cover.

    Task completed _____

STUDENT SELF-EVALUATION

Check	Level	Competency	Comments
	4	Mastered task	
	3	Competent but need further help	
	2	Needed a lot of help	
	0	Did not understand the task	

INSTRUCTOR EVALUATION

Check	Level	Competency	Comments
	4	Mastered task	
	3	Competent but needs further help	
	2	Requires more training	
	0	Unable to perform task	

ONLINE TASKS

Use a search engine and locate the Insurance Institute for Highway Safety (IIHS) Web site. Using its links, check out what the IIHS has to say about hydraulic brakes versus air brakes in truck applications. Also check out its Q&A forms on other vehicle safety issues.

STUDY TIPS

Identify 5 key points in Chapter 29. Try to be as brief as possible.

Key point 1 _____

Key point 2 _____

Key point 3 _____

Key point 4 _____

Key point 5 _____

30 ABS and EBS

Objectives

After reading this chapter, you should be able to:
- Describe how an antilock brake system (ABS) works to prevent wheel lockup during braking.
- List the major components of a truck ABS.
- Describe the operation of ABS input circuit components.
- Outline the role of the ABS module when managing antiskid mode.
- Explain how the ABS module controls the service modulator valves.
- Explain what is meant by the number of channels of an ABS.
- Describe how trailer ABS is managed.
- Outline the procedure for diagnosing ABS faults.
- Describe the procedure required to set up and adjust a wheel speed sensor.
- Explain how ABS, EBS, AEB, and ADAS are used to enable semi- and fully autonomous truck operation.
- Explain how an electronic brake system (EBS) manages service brake applications.
- Describe some ABS add-ons such as stability control electronics.
- Outline the reasons why an EBS has to meet current FMVSS 121 requirements.

PRACTICE QUESTIONS

1. What happens if an ABS ceases to function?
   a. An LED flashes a code directing the driver when to apply the brakes.
   b. The system reverts to normal non-ABS operation.
   c. A fail-safe device shifts to a secondary ABS.
   d. The entire braking system shuts down.

2. Which type of sensor do pneumatic ABS brakes use to signal wheel speed data to the electronic control unit (ECU)?
   a. variable capacitance
   b. pulse wheel
   c. potentiometer
   d. hall effect

3. When an ABS is first powered up, what causes the clicking noise?
   a. brake shoes contacting the drums during the self-test sequence
   b. the ECU booting up
   c. solenoids during the self-test sequence
   d. air exhausted by control valves

4. Technician A says that in a combination tractor/semi-trailer rig, both equipped with ABS, the ECU on the tractor controls ABS on the trailer. Technician B says that today all new highway trucks and trailers are required to have ABS. Who is correct?
   a. Technician A only
   b. Technician B only
   c. both A and B
   d. neither A nor B

5. In a tandem drive tractor with 4S/4M air ABS, which of the following would be true?
   a. Traction control is provided.
   b. Better split-coefficient braking performance is provided than with the 6S/6M air ABS.
   c. Steering axle brakes are modulated as a pair.
   d. The drive axle brakes are modulated in pairs.

6. Which of the following accounts for the lower modulation rates of air brake ABS when compared with hydraulic ABS?
   a. heavier foundation brake components
   b. compressibility of air
   c. lower efficiency of S-cam brakes
   d. greater heat generated

7. What type of voltage signal is produced by an ABS wheel speed sensor?
   a. AC
   b. PWM
   c. DC negative
   d. DC positive

8. How many teeth does a current truck ABS wheel speed pulse (chopper) wheel have?
   a. 1
   b. 16
   c. 32
   d. 100

9. What is the lowest threshold voltage value at which ABS will cease to function properly?
   a. 9.6 volts
   b. 11 volts
   c. 12.6 volts
   d. 14 volts

10. When an EBS is used on a truck air brake system, which of the following circuits is replaced by an electronic circuit?
    a. supply circuit
    b. control circuit
    c. service circuit
    d. park circuit

JOB SHEET 30.1

Name _____ Station _____ Date _____

Verify the Operation of a Wheel Speed Sensor.

Performance Objective(s): Check out the operation of a wheel speed sensor using a lab scope, and adjust the gap if required.

ASE Education Foundation Correlation

This job sheet addresses the following ASE Education Foundation task(s):

III. Brakes
   A. General
   1. Research vehicle service information, including fluid type, vehicle service history, service precautions, and technical service bulletins. (IMMR, TST, MTST) P-1.
   2. Identify brake system components and configurations (including air and hydraulic systems, parking brake, power assist, and vehicle dynamic brake systems). (IMMR, TST, MTST) P-1.

   I. Vehicle Dynamic Brake Systems (Air and Hydraulic): Antilock Brake System (ABS), Automatic Traction Control (ATC) System, and Electronic Stability Control (ESC) System
   1. Observe antilock brake system (ABS) warning light operation including trailer and dash-mounted trailer ABS warning light. (IMMR) P-1.
   1. Observe antilock brake system (ABS) warning light operation including trailer and dash-mounted trailer ABS warning light; determine needed action. (TST, MTST) P-1.
   2. Observe automatic traction control (ATC) and electronic stability control (ESC) warning light operation. (IMMR) P-2.
   2. Observe automatic traction control (ATC) and electronic stability control (ETC) warning light operation; determine needed action. (TST, MTST) P-2.
   3. Identify stopping concerns related to the vehicle dynamic brake systems: ABS, ATC, and ESC; determine needed action. (TST, MTST) P-2.
   6. Test vehicle/wheel speed sensors and circuits; adjust, repair, and/or replace as needed. (TST, MTST) P-1.
   7. Bleed ABS hydraulic circuits. (TST, MTST) P-2.

Tools and Materials: A truck equipped with ABS, a lab scope, a DMM with frequency display, thickness gauges, and OEM service literature

Protective Clothing: Standard shop apparel including coveralls or shop coat, safety glasses, and safety footwear

PROCEDURE
Ensure that the shop LOTO is observed.

1. Raise one wheel of the vehicle to be tested, ensuring that it has a wheel speed sensor input to the ABS ECU. Place the axle on a stand.

   Task completed _____

2. Turn on the ignition circuit.

   Task completed _____

3. Visually inspect the wheel speed reluctor and pickup for evidence of damage.

   Task completed _____

4. Visually inspect the wiring and connectors from the wheel speed sensor and the ECU.

   Task completed _____

5. Connect a lab scope across the wheel sensor.

   Task completed _____

6. Rotate the wheel by hand and observe the waveform displayed on the lab scope. As the wheel turns, the waveform should produce nodes and antinodes above and below 0 AC volts. The waveforms should increase

in frequency with increased rotational speed. Record the AC voltage produced when the wheel is rotated by hand at approximately 30 rpm and check to OEM specification. Also check the resistance of the sensor: some OEMs only provide a resistance specification for wheel speed sensors.

Task completed _____

7. Check the air gap specification for the wheel sensor used.

Task completed _____

8. Use a set of thickness gauges to measure the gap between the pickup and reluctor teeth on the rotor. Check in four locations and record the mean gap dimension.

Task completed _____

9. Adjust the sensor gap if out of specification. Consult the OEM instructions; in some instances, this may require the replacement of the pickup.

Task completed _____

STUDENT SELF-EVALUATION

Check	Level	Competency	Comments
	4	Mastered task	
	3	Competent but need further help	
	2	Needed a lot of help	
	0	Did not understand the task	

INSTRUCTOR EVALUATION

Check	Level	Competency	Comments
	4	Mastered task	
	3	Competent but needs further help	
	2	Requires more training	
	0	Unable to perform task	

ONLINE TASKS

Use a search engine to locate the Insurance Institute for Highway Safety (IIHS) Web site and, using its information, discuss whether mandatory ABS on trucks has resulted in making our roads safer.

STUDY TIPS

Identify 5 key points in Chapter 30. Try to be as brief as possible.

Key point 1 _____

Key point 2 _____

Key point 3 _____

Key point 4 _____

Key point 5 _____

31 Air Brake Servicing

Objectives

After reading this chapter, you should be able to:
- Understand the safety requirements of working on an air brake system.
- Perform basic maintenance on an air brake system.
- Diagnose common compressor problems.
- Describe the procedure required to service an air dryer.
- Performance test an air dryer.
- Check out the service brakes on a truck.
- Test the emergency and parking brake systems.
- Verify the operation of the trailer brakes.
- Understand the OOS criteria used by safety inspection officers.
- Diagnose some brake valve failures.
- Describe the procedure required to overhaul foundation brakes.
- Determine brake freestroke and identify when an adjustment is required.
- Recognize different types of brake stroke indicator and identify an overstroke condition.
- Outline some common service procedures used on air disc brake systems.
- Outline some of the changes that have taken place in commercial tractor/trailer air brake systems since the introduction in 2010 of FMVSS 121 revised stopping distances.

PRACTICE QUESTIONS

1. Technician A drains the air reservoirs weekly. Technician B says that spring brake chambers are designed to apply the parking brakes in the event of loss of pressure in the hold-off chambers. Who is correct?
   a. Technician A only
   b. Technician B only
   c. both A and B
   d. neither A nor B

2. In an air brake system, the FMVSS 121 standard for pressure build from 85 to 100 psi should be within:
   a. 15 seconds
   b. 25 seconds
   c. 1 minute
   d. 2 minutes

3. Technician A says that timing compatibility is ensured when tractors and trailers meet FMVSS No. 121 application and release timing requirements. Technician B says that the vast majority of brake applications are made at less than 30 psi. Who is correct?
   a. Technician A only
   b. Technician B only
   c. both A and B
   d. neither A nor B

4. A truck air compressor is under effective cycle almost continuously and even then fails to maintain sufficient pressure. Technician A says that the problem could be a dirty intake cleaner. Technician B says that it could be a leaking exhaust valve. Who is correct?
   a. Technician A only
   b. Technician B only
   c. both A and B
   d. neither A nor B

5. Technician A says that it is recommended practice to disassemble, clean, and inspect the air governor every 100,000 miles (160,000 km). Technician B says that a compressor should be overhauled every 100,000 miles (160,000 km). Who is correct?
   a. Technician A only
   b. Technician B only
   c. both A and B
   d. neither A nor B

6. The purge cycle of an air dryer lasts more than 40 seconds. Technician A says that the one-way check valve may be defective. Technician B says that the purge valve may be stuck. Who is correct?
   a. Technician A only
   b. Technician B only
   c. both A and B
   d. neither A nor B

7. Which of the following correctly describes the operation of a quick-release valve?
   a. exhausts signal pressure at the valve
   b. exhausts downstream pressure at the valve
   c. distributed pressure is always less than signal pressure
   d. all of the above

8. Technician A says that some types of spring brake valves are designed to regulate the hold-off pressure. Technician B says that a quick-release valve can act to speed up brake release times. Who is correct?
   a. Technician A only
   b. Technician B only
   c. both A and B
   d. neither A nor B

9. Technician A says that the difference between air governor cut-in and cut-out pressures can be adjusted with the governor adjusting screw. Technician B says that a cause of frequent cycling of the compressor is air leakage. Who is correct?
   a. Technician A only
   b. Technician B only
   c. both A and B
   d. neither A nor B

10. Air pressure builds in the system far too slowly. This could be caused by any of the following, except:
    a. a clogged compressor air strainer
    b. worn compressor piston rings
    c. a frozen line to the governor
    d. excessive carbon in the compressor

JOB SHEET 31.1

Name _____ Station _____ Date _____

Performance Check the Supply and Control Circuits of an Air Brake System.

Performance Objective(s): Learn the procedure required to verify the performance of the control and supply circuits of a tractor/trailer air brake system.

ASE Education Foundation Correlation

This job sheet addresses the following ASE Education Foundation task(s):

III. Brakes
  A. General
   1. Research vehicle service information, including fluid type, vehicle service history, service precautions, and technical service bulletins. (IMMR, TST, MTST) P-1.
   2. Identify brake system components and configurations (including air and hydraulic systems, parking brake, power assist, and vehicle dynamic brake systems). (IMMR, TST, MTST) P-1.
   3. Identify brake performance problems caused by the mechanical/foundation brake system (air and hydraulic). (IMMR, TST, MTST) P-1.
   4. Use appropriate electronic service tool(s) and procedures to diagnose problems; check, record, and clear diagnostic codes; interpret digital multimeter (DMM) readings. (TST, MTST) P-1.

  B. Air Brakes: Air Supply and Service Systems
   1. Inspect air supply system components such as compressor, governor, air drier, tanks, and lines; inspect service system components such as lines, fittings, mountings, and valves (hand brake/trailer control, brake relay, quick release, tractor protection, emergency/spring brake control/modulator, pressure relief/safety). (IMMR) P-1.
   1. Inspect, test, repair, and/or replace air supply system components such as compressor, governor, air drier, tanks, and lines; inspect service system components such as lines, fittings, mountings, and valves (hand brake/trailer control, brake relay, quick release, tractor protection, emergency/spring brake control/modulator, pressure relief/safety); determine needed action. (TST, MTST) P-1.
   2. Verify proper gauge operation and readings; verify low pressure warning alarm operation; perform air supply system tests such as pressure build-up, governor settings, and leakage; drain air tanks and check for contamination. (IMMR) P-1.
   2. Test gauge operation and readings; test low pressure warning alarm operation; perform air supply system tests such as pressure build-up, governor settings, and leakage; drain air tanks and check for contamination; determine needed action. (TST, MTST) P-1.
   3. Demonstrate knowledge and understanding of air supply and service system components and operations. (TST, MTST) P-1.
   4. Inspect air compressor drive gear components (gears, belts, tensioners, and/or couplings); determine needed action. (TST, MTST) P-3.
   5. Inspect air compressor inlet; inspect oil supply and coolant lines, fittings, and mounting brackets; repair or replace as needed. (TST, MTST) P-1.
   6. Inspect and test air tank relief (safety) valves, one-way (single) check valves, two-way (double) check valves, manual and automatic drain valves; determine needed action. (TST, MTST) P-1.
   7. Inspect and clean air drier systems, filters, valves, heaters, wiring, and connectors; determine needed action. (TST, MTST) P-1.
   8. Inspect and test brake application (foot/treadle) valve, fittings, and mounts; check pedal operation; determine needed action. (TST, MTST) P-1.

  C. Air Brakes: Mechanical/Foundation Brake System
   1. Inspect service brake chambers, diaphragms, clamps, springs, pushrods, clevises, and mounting brackets; determine needed action. (IMMR) P-1.
   1. Inspect and test service brake chambers, diaphragms, clamps, springs, pushrods, clevises, and mounting brackets; determine needed action. (TST) P-1.
   1. Inspect, test, repair, and/or replace service brake chambers, diaphragms, clamps, springs, pushrods, clevises, and mounting brackets; determine needed action. (MTST) P-1.

2. Identify slack adjuster type; inspect slack adjusters; determine needed action. (IMMR) P-1.
2. Identify slack adjuster type; inspect slack adjusters; perform needed action. (TST, MTST) P-1.
3. Check camshafts (S-cams), tubes, rollers, bushings, seals, spacers, retainers, brake spiders, shields, anchor pins, and springs; determine needed action. (IMMR) P-1.
3. Check camshafts (S-cam), tubes, rollers, bushings, seals, spacers, retainers, brake spiders, shields, anchor pins, and springs; perform needed action. (TST, MTST) P-1.
4. Inspect rotor and mounting surface; measure rotor thickness, thickness variation, and lateral runout; determine needed action. (IMMR, TST, MTST) P-1.
5. Inspect, clean, and adjust air disc brake caliper assemblies; inspect and measure disc brake pads; inspect mounting hardware; perform needed action. (IMMR, TST, MTST) P-1.
6. Remove brake drum; clean and inspect brake drum and mounting surface; measure brake drum diameter; measure brake lining thickness; inspect brake lining condition; determine needed action. (IMMR, TST, MTST) P-1.
7. Identify concerns related to the mechanical/foundation brake system including poor stopping, brake noise, premature wear, pulling, grabbing, or dragging; determine needed action. (TST) P-1.
7. Diagnose concerns related to the mechanical/foundation brake system including poor stopping, brake noise, premature wear, pulling, grabbing, or dragging; determine needed action. (MTST) P-1.

D. Air Brakes: Parking Brake System
1. Inspect and check parking (spring) brake chamber for leaks; determine needed action. (IMMR) P-1.
1. Inspect and test service brake chambers, diaphragms, clamps, springs, pushrods, clevises, and mounting brackets; determine needed action. (TST) P-1.
1. Inspect, test, and/or replace parking (spring) brake chamber. (MTST) P-1.
2. Inspect and test parking (spring) brake check valves, lines, hoses, and fittings; determine needed action. (IMMR) P-1.
2. Inspect, test, and/or replace parking (spring) brake check valves, lines, hoses, and fittings. (TST, MTST) P-1.
3. Inspect and test parking (spring) brake application and release valve; determine needed action. (IMMR) P-1.
3. Inspect, test, and/or replace parking (spring) brake application and release valve. (TST, MTST) P-1.
4. Manually release (cage) and reset (uncage) parking (spring) brakes. (IMMR, TST, MTST) P-1.
5. Identify and test anti compounding brake function; determine needed action. (TST, MTST) P-2.

F. Hydraulic Brakes: Mechanical/Foundation Brake System
1. Inspect rotor and mounting surface; measure rotor thickness, thickness variation, and lateral runout; determine needed action. (IMMR, TST) P-1.
1. Clean and inspect rotor and mounting surface; measure rotor thickness, thickness variation, and lateral runout; determine necessary action. (MTST) P-1.
2. Inspect and clean disc brake caliper assemblies; inspect and measure disc brake pads; inspect mounting hardware; determine needed action. (IMMR) P-1.
2. Inspect and clean disc brake caliper assemblies; inspect and measure disc brake pads; inspect mounting hardware; perform needed action. (TST, MTST) P-1.
3. Remove brake drum; clean and inspect brake drum and mounting surface; measure brake drum diameter; measure brake lining thickness; inspect brake lining condition; inspect wheel cylinders; determine needed action. (IMMR, TST, MTST) P-1.

H. Power Assist Systems
1. Check brake assist/booster system (vacuum or hydraulic) hoses and control valves; check fluid level and condition (if applicable). (IMMR, TST, MTST) P-1.

I. Vehicle Dynamic Brake Systems (Air and Hydraulic): Antilock Brake System (ABS), Automatic Traction Control (ATC) System, and Electronic Stability Control (ESC) System
1. Observe antilock brake system (ABS) warning light operation including trailer and dash mounted trailer ABS warning light. (IMMR) P-1.
1. Observe antilock brake system (ABS) warning light operation including trailer and dash mounted trailer ABS warning light; determine needed action. (TST, MTST) P-1.
2. Observe automatic traction control (ATC) and electronic stability control (ESC) warning light operation. (IMMR) P-2.
2. Observe automatic traction control (ATC) and electronic stability control (ETC) warning light operation; determine needed action. (TST, MTST) P-2.
3. Identify stopping concerns related to the vehicle dynamic brake systems: ABS, ATC, and ESC; determine needed action. (TST, MTST) P-2.
6. Test vehicle/wheel speed sensors and circuits; adjust, repair, and/or replace as needed. (TST, MTST) P-1.
8. Verify power line carrier (PLC) operation. (TST, MTST) P-3.

J. Wheel Bearings
1. Clean, inspect, lubricate, and/or replace wheel bearings and races/cups; replace seals and wear rings; inspect spindle/tube; inspect and replace retaining hardware; adjust wheel bearings; check hub assembly fluid level and condition; verify end play with dial indicator method. (IMMR, TST, MTST) P-1.
2. Identify, inspect, and/or replace unitized/preset hub bearing assemblies. (IMMR, TST, MTST) P-2.

Tools and Materials: A coupled tractor/trailer rig with a functioning air brake system, and standard shop hand and power tools

Protective Clothing: Standard shop apparel including coveralls or shop coat, safety glasses, and safety footwear

PROCEDURE
Ensure that the shop LOTO is observed.

1. Park the vehicle and block one set of wheels on the tractor and one set on the trailer.

   Task completed _____

2. Check the primary and secondary reservoir check valves for correct operation by following this sequence:

   a. Build air pressure to system pressure.

   Task completed _____

   b. With the ignition switch on, open the drain valve at the supply (wet) reservoir and completely drain.

   Task completed _____

   c. Does the low air pressure buzzer come on at approximately 70 psi?

   Yes_____ No_____

   Task completed _____

   d. Do both the primary and secondary reservoirs retain full system air pressure?

   Yes_____ No_____

   Task completed _____

3. Actuate the System Park control (yellow) valve and the Trailer Supply control (red) valve by pushing in to the released position.

   Task completed _____

4. Now open the drain valves in the primary and secondary reservoirs.

   Task completed _____

   a. Does the red knob on the tractor supply control valve pop out (apply position) when the reservoir with the highest pressure reaches 40 ± 6 psi?

   Yes_____ No_____

   Task completed _____

   b. Does the yellow knob on the park brake control valve pop out when the reservoir with the higher pressure reaches 30 ± 5 psi?

   Yes_____ No_____

   Task completed _____

5. Close all reservoir draincocks.

   Task completed _____

6. Build up air supply in the chassis system to system pressure.

   Task completed _____

7. Actuate the Trailer Supply (red) knob to release the trailer parking brakes.

   Task completed _____

8. Disconnect the trailer supply gladhand.

   Task completed _____

   a. Does the dash trailer supply valve (red) knob pop out instantly?

      Yes_____ No_____

      Task completed _____

   b. If not, proceed to the following steps.

      Task completed _____

9. Reconnect the trailer supply gladhand.

   Task completed _____

10. Push the System Park (yellow) knob dash valve to the released position (in).

    Task completed _____

11. Pull the Trailer Supply knob to the applied (park) position (out).

    Task completed _____

12. Check the air system for leakage by observing the air gauges on the instrument panel.

    Task completed _____

    Is leakage greater than 2 psi in 1 minute?

    Yes_____ No_____

    Task completed _____

13. Open the draincock in the secondary air reservoir.

    Task completed _____

    a. Observe the air gauges to detect loss of air.

       Task completed _____

    b. Does the system show a loss of air?

       Yes_____ No_____

       Task completed _____

    c. Does the low-pressure indicator buzzer in the cab sound at approximately 70 psi when the ignition switch is on?

       Yes_____ No_____

       Task completed _____

14. Ensure that the parking brakes are released and apply the service brakes. Observe the slack adjusters and service brake chamber pushrods. Check and record which brakes apply.

    Task completed _____

15. Close the draincock in the secondary air reservoir.

    Task completed _____

16. Build the air supply in the air system to system pressure.

    Task completed _____

17. Open the draincock in the primary reservoir.

    Task completed _____

18. Observe the secondary circuit gauge for loss of air.

    Task completed _____

19. Apply the service brakes and observe the slack adjusters and service brake chamber pushrods. Check the brakes that apply and note those that do not.

Task completed _____

20. Close all draincocks and build pressure to system pressure.

Task completed _____

Review Figure 31–1, Figure 31–2, Figure 31–3, Figure 31–4, Figure 31–5, and Figure 31–6, which show air brake diagnostic equipment and how to locate the various test devices into the air brake circuit. Attempt to perform these tests on both functional and malfunctioning air brake circuits.

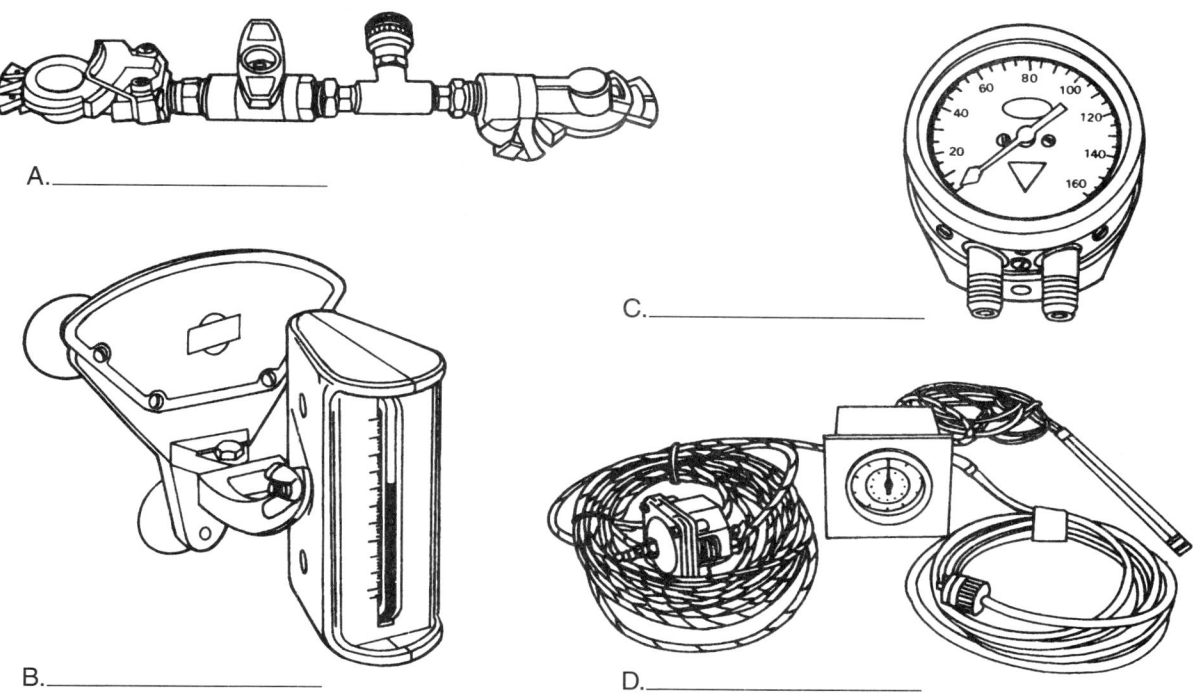

A. _____

B. _____

C. _____

D. _____

FIGURE 31–1 Identify by filling in the blanks and describe the function of these air brake test components.

Air Brake Servicing 327

TORQUE
BALANCE TEST
1
3
4
5
Trailer
service
Trailer
emergency
Tractor
2
Trailer
PRESSURE BALANCE AND
PRESSURE BUILDUP RATE
TIMING TEST
1 Duplex gauge or single gauge
2 Duplex gauge
3 Double gladhand assembly
4 Park brake control valve
5 Trailer supply control valve

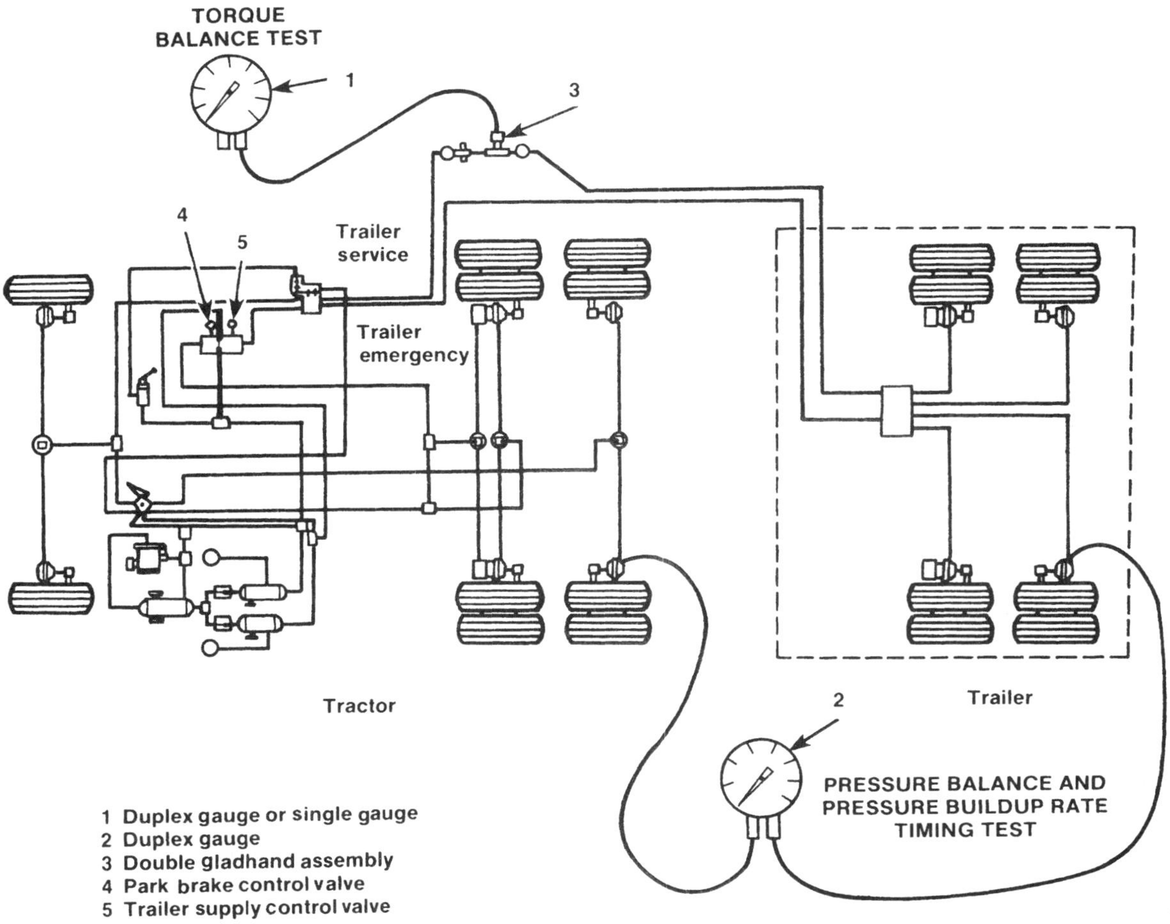

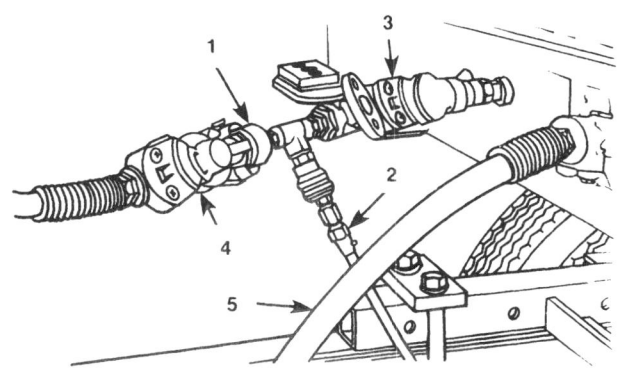

FIGURE 31–2 Test setup for torque and pressure balance diagnosis of a single trailer system

1
2
3
4
5
1 Double Gladhand Assembly
2 To Duplex Air Gauge
3 Trailer Service Brake Connection
4 Tractor Service Brake Connection
5 Tractor Emergency Brake Connection

FIGURE 31–3 Installation location of a double gladhand assembly

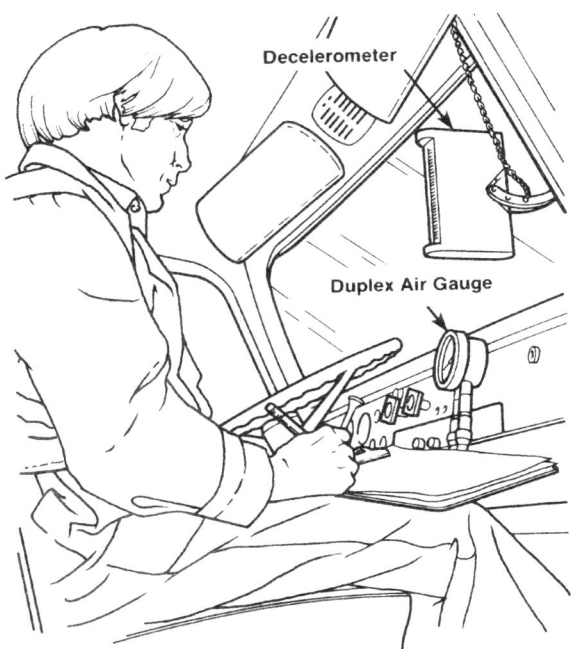

FIGURE 31-4 Location of test instrumentation in the cab

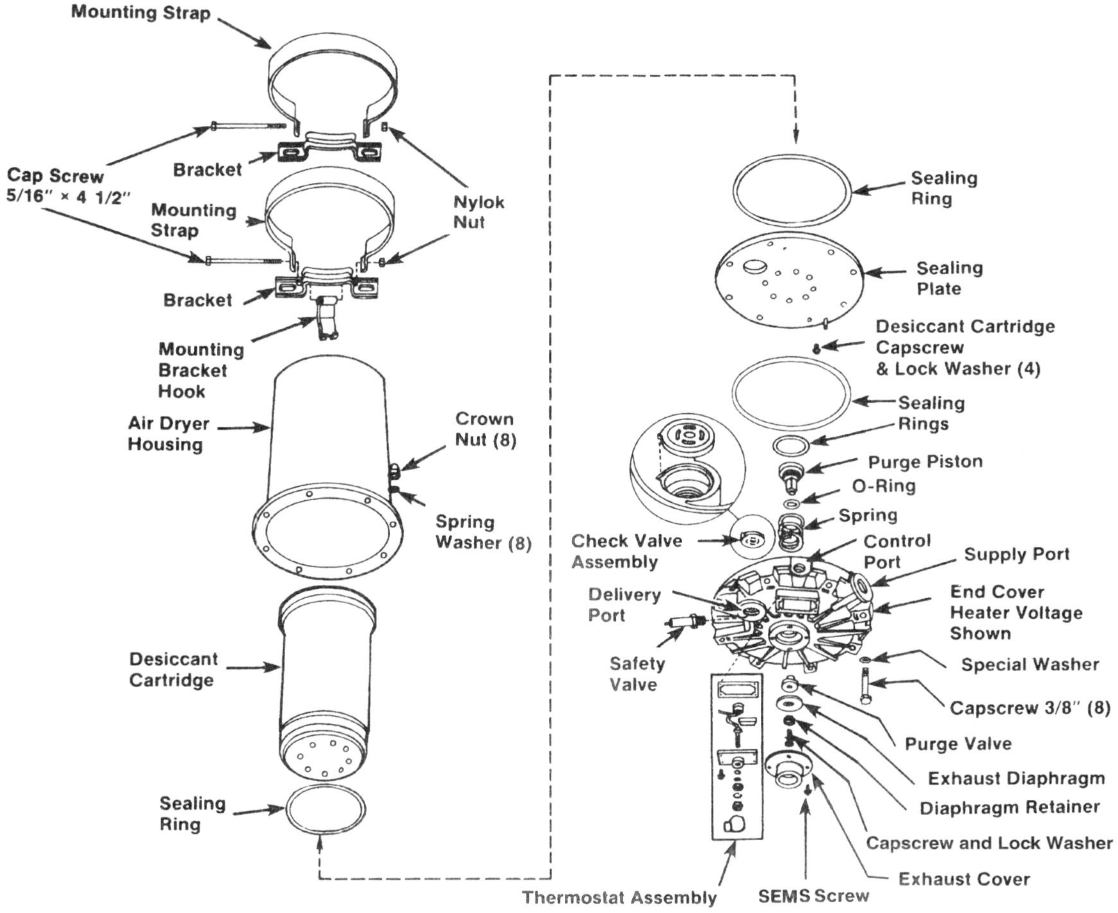

FIGURE 31-5 Exploded view of an air dryer assembly

Air Brake Servicing 329

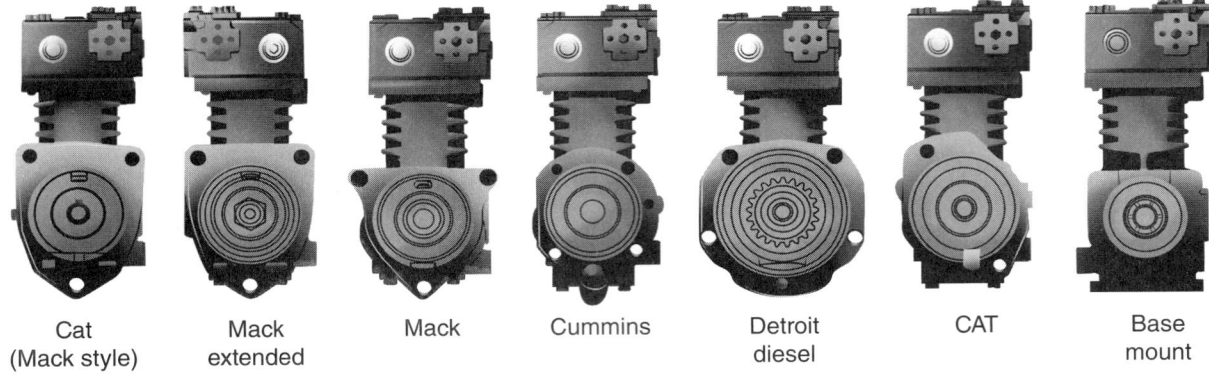

FIGURE 31-6 Compressor mountings required by different engine manufacturers

STUDENT SELF-EVALUATION

Check	Level	Competency	Comments
	4	Mastered task	
	3	Competent but need further help	
	2	Needed a lot of help	
	0	Did not understand the task	

INSTRUCTOR EVALUATION

Check	Level	Competency	Comments
	4	Mastered task	
	3	Competent but needs further help	
	2	Requires more training	
	0	Unable to perform task	

JOB SHEET 31.2

Name _____ Station _____ Date _____

Perform an FMVSS 121 Brake Test Using an OEM (Bendix) Test Profile.

Performance Objective(s): Perform an FMVSS 121 brake test using an OEM test profile to verify the performance of a chassis air brake system.

ASE Education Foundation Correlation

This job sheet addresses the following ASE Education Foundation task(s):

III. Brakes
   A. General
   1. Research vehicle service information, including fluid type, vehicle service history, service precautions, and technical service bulletins. (IMMR, TST, MTST) P-1.
   2. Identify brake system components and configurations (including air and hydraulic systems, parking brake, power assist, and vehicle dynamic brake systems). (IMMR, TST, MTST) P-1.
   3. Identify brake performance problems caused by the mechanical/foundation brake system (air and hydraulic). (IMMR, TST, MTST) P-1.
   4. Use appropriate electronic service tool(s) and procedures to diagnose problems; check, record, and clear diagnostic codes; interpret digital multimeter (DMM) readings. (TST, MTST) P-1.

   B. Air Brakes: Air Supply and Service Systems
   1. Inspect air supply system components such as compressor, governor, air drier, tanks, and lines; inspect service system components such as lines, fittings, mountings, and valves (hand brake/trailer control, brake relay, quick release, tractor protection, emergency/spring brake control/modulator, pressure relief/safety). (IMMR) P-1.
   1. Inspect, test, repair, and/or replace air supply system components such as compressor, governor, air drier, tanks, and lines; inspect service system components such as lines, fittings, mountings, and valves (hand brake/trailer control, brake relay, quick release, tractor protection, emergency/spring brake control/modulator, pressure relief/safety); determine needed action. (TST, MTST) P-1.
   2. Verify proper gauge operation and readings; verify low pressure warning alarm operation; perform air supply system tests such as pressure build-up, governor settings, and leakage; drain air tanks and check for contamination. (IMMR) P-1.
   2. Test gauge operation and readings; test low pressure warning alarm operation; perform air supply system tests such as pressure build-up, governor settings, and leakage; drain air tanks and check for contamination; determine needed action. (TST, MTST) P-1.
   3. Demonstrate knowledge and understanding of air supply and service system components and operations. (TST, MTST) P-1.
   4. Inspect air compressor drive gear components (gears, belts, tensioners, and/or couplings); determine needed action. (TST, MTST) P-3.
   5. Inspect air compressor inlet; inspect oil supply and coolant lines, fittings, and mounting brackets; repair or replace as needed. (TST, MTST) P-1.
   6. Inspect and test air tank relief (safety) valves, one-way (single) check valves, two-way (double) check valves, manual and automatic drain valves; determine needed action. (TST, MTST) P-1.
   7. Inspect and clean air drier systems, filters, valves, heaters, wiring, and connectors; determine needed action. (TST, MTST) P-1.
   8. Inspect and test brake application (foot/treadle) valve, fittings, and mounts; check pedal operation; determine needed action. (TST, MTST) P-1.

   C. Air Brakes: Mechanical/Foundation Brake System
   1. Inspect service brake chambers, diaphragms, clamps, springs, pushrods, clevises, and mounting brackets; determine needed action. (IMMR) P-1.
   1. Inspect and test service brake chambers, diaphragms, clamps, springs, pushrods, clevises, and mounting brackets; determine needed action. (TST) P-1.
   1. Inspect, test, repair, and/or replace service brake chambers, diaphragms, clamps, springs, pushrods, clevises, and mounting brackets; determine needed action. (MTST) P-1.

2. Identify slack adjuster type; inspect slack adjusters; determine needed action. (IMMR) P-1.
2. Identify slack adjuster type; inspect slack adjusters; perform needed action. (TST, MTST) P-1.
3. Check camshafts (S-cams), tubes, rollers, bushings, seals, spacers, retainers, brake spiders, shields, anchor pins, and springs; determine needed action. (IMMR) P-1.
3. Check camshafts (S-cam), tubes, rollers, bushings, seals, spacers, retainers, brake spiders, shields, anchor pins, and springs; perform needed action. (TST, MTST) P-1.
4. Inspect rotor and mounting surface; measure rotor thickness, thickness variation, and lateral runout; determine needed action. (IMMR, TST, MTST) P-1.
5. Inspect, clean, and adjust air disc brake caliper assemblies; inspect and measure disc brake pads; inspect mounting hardware; perform needed action. (IMMR, TST, MTST) P-1.
6. Remove brake drum; clean and inspect brake drum and mounting surface; measure brake drum diameter; measure brake lining thickness; inspect brake lining condition; determine needed action. (IMMR, TST, MTST) P-1.
7. Identify concerns related to the mechanical/foundation brake system including poor stopping, brake noise, premature wear, pulling, grabbing, or dragging; determine needed action. (TST) P-1.
7. Diagnose concerns related to the mechanical/foundation brake system including poor stopping, brake noise, premature wear, pulling, grabbing, or dragging; determine needed action. (MTST) P-1.

D. Air Brakes: Parking Brake System
1. Inspect and check parking (spring) brake chamber for leaks; determine needed action. (IMMR) P-1.
1. Inspect and test service brake chambers, diaphragms, clamps, springs, pushrods, clevises, and mounting brackets; determine needed action. (TST) P-1.
1. Inspect, test, and/or replace parking (spring) brake chamber. (MTST) P-1.
2. Inspect and test parking (spring) brake check valves, lines, hoses, and fittings; determine needed action. (IMMR) P-1.
2. Inspect, test, and/or replace parking (spring) brake check valves, lines, hoses, and fittings. (TST, MTST) P-1.
3. Inspect and test parking (spring) brake application and release valve; determine needed action. (IMMR) P-1.
3. Inspect, test, and/or replace parking (spring) brake application and release valve. (TST, MTST) P-1.
4. Manually release (cage) and reset (uncage) parking (spring) brakes. (IMMR, TST, MTST) P-1.
5. Identify and test anti compounding brake function; determine needed action. (TST, MTST) P-2.

F. Hydraulic Brakes: Mechanical/Foundation Brake System
1. Inspect rotor and mounting surface; measure rotor thickness, thickness variation, and lateral runout; determine needed action. (IMMR, TST) P-1.
1. Clean and inspect rotor and mounting surface; measure rotor thickness, thickness variation, and lateral runout; determine necessary action. (MTST) P-1.
2. Inspect and clean disc brake caliper assemblies; inspect and measure disc brake pads; inspect mounting hardware; determine needed action. (IMMR) P-1.
2. Inspect and clean disc brake caliper assemblies; inspect and measure disc brake pads; inspect mounting hardware; perform needed action. (TST, MTST) P-1.
3. Remove brake drum; clean and inspect brake drum and mounting surface; measure brake drum diameter; measure brake lining thickness; inspect brake lining condition; inspect wheel cylinders; determine needed action. (IMMR, TST, MTST) P-1.

H. Power Assist Systems
1. Check brake assist/booster system (vacuum or hydraulic) hoses and control valves; check fluid level and condition (if applicable). (IMMR, TST, MTST) P-1.

I. Vehicle Dynamic Brake Systems (Air and Hydraulic): Antilock Brake System (ABS), Automatic Traction Control (ATC) System, and Electronic Stability Control (ESC) System
1. Observe antilock brake system (ABS) warning light operation including trailer and dash-mounted trailer ABS warning light. (IMMR) P-1.
1. Observe antilock brake system (ABS) warning light operation including trailer and dash-mounted trailer ABS warning light; determine needed action. (TST, MTST) P-1.
2. Observe automatic traction control (ATC) and electronic stability control (ESC) warning light operation. (IMMR) P-2.
2. Observe automatic traction control (ATC) and electronic stability control (ETC) warning light operation; determine needed action. (TST, MTST) P-2.

3. Identify stopping concerns related to the vehicle dynamic brake systems: ABS, ATC, and ESC; determine needed action. (TST, MTST) P-2.
5. Check and test operation of vehicle dynamic brake system (air and hydraulic) mechanical and electrical components; determine needed action. (TST, MTST) P-1.
6. Test vehicle/wheel speed sensors and circuits; adjust, repair, and/or replace as needed. (TST, MTST) P-1.
7. Bleed ABS hydraulic circuits. (TST, MTST) P-1.
8. Verify power line carrier (PLC) operation. (TST, MTST) P-3.

J. Wheel Bearings
1. Clean, inspect, lubricate, and/or replace wheel bearings and races/cups; replace seals and wear rings; inspect spindle/tube; inspect and replace retaining hardware; adjust wheel bearings; check hub assembly fluid level and condition; verify end play with dial indicator method. (IMMR, TST, MTST) P-1.
2. Identify, inspect, and/or replace unitized/preset hub bearing assemblies. (IMMR, TST, MTST) P-2.

Tools and Materials: A coupled tractor/trailer rig with a functioning air brake system and standard shop hand and power tools

Protective Clothing: Standard shop apparel including coveralls or shop coat, safety glasses, and safety footwear

PROCEDURE
Ensure that the shop LOTO is observed.

Complete the test profile that follows and make any required repairs.

FMVSS 121 Dual Air Brake System

Test and Checklist

Date: _____

Truck: _____ Mileage:_____

Trailer:_____ Mileage:_____

Technician:_____

Prechecks:

1. Examine all tubing for signs of kinks or dents.

2. Examine all hoses for signs of wear, drying out, or overheating.

3. Check the suspension of all tubing. It should be supported and not vibrate.

4. Check the suspension of all hoses. Position it so that the hose will not abrade or be subject to excessive heat.

Test 1
GOVERNOR CUT-OUT / LOW PRESSURE WARNING / PRESSURE BUILD-UP

Vehicle Parked, Wheel Chocked OK NOT OK

1. Drain all reservoirs to 0 psi
2. Start engine (run at fast idle): Low pressure warning should be on. Note: On some vehicles with antilock, warning light will also come on momentarily when ignition is turned on.
3. Low pressure warning: Dash warning light should go off at or above 60 psi.
4. Build up time: Pressure should build from 85 to 100 psi within 25 seconds.
5. Governor cut-out: Cuts out at correct pressure. Check manufacturer's recommendations; usually 100–130 psi.
6. Governor cut-in: Reduce service air pressure to governor cut-in. The difference between cut-in and cut-out pressure must not exceed 25 psi.

COMPLETE NECESSARY REPAIRS BEFORE PROCEEDING TO TEST 2; SEE CHECKLIST 1 FOR COMMON CORRECTIONS.

Check List 1

If the low pressure warning light or buzzer doesn't come on:
1. Check wiring
2. Check bulb
3. Repair or replace the buzzer, bulb, or low pressure warning switch(es)

If governor cut-out is higher or lower than specified by the vehicle manual:
1. Adjust the governor using a gauge of known accuracy.
2. Repair or replace governor as necessary after being sure compressor unloader mechanism is operating correctly.

If low pressure warning occurs below 60 psi:
1. Check dash gauge with master pressure gauge.
2. Repair or replace a defective low pressure indicator.

If build up time exceeds 25 seconds:
1. Test the compressor air filter and clean or replace.
2. Check for restricted inlet line if compressor does not have a filter. Replace if necessary.
3. Check compressor discharge port and line for carbon build-up. Replace if necessary.
4. With system charged and governor compressor in unloaded mode, listen at the compressor inlet for leakage. If leakage is evident, apply a small quantity of oil around unloader pistons. If no leakage is evident, then leakage is through the compressor discharge valves.
5. Check the compressor drive for slippage.

RETEST TO CHECK OUT ALL ITEMS REPAIRED OR REPLACED

Test 2
LEAKAGE AT RESERVOIR AIR SUPPLY

GOVERNOR CUT-OUT PRESSURE, ENGINE OFF
PARKING BRAKES APPLIED   OK   NOT OK

1. Allow pressure to stabilize for 1 minute   ☐   ☐
2. Observe the dash gauge pressures for 2 minutes and note any pressure drop
   - A. Pressure Drop: Single Vehicle (A 2 psi drop within 2 minutes is permissible for either service reservoir)   ☐   ☐
   - B. Pressure Drop: Tractor/Trailer (A 6 psi drop within 2 minutes is permissible for either service reservoir)   ☐   ☐
   - C. Pressure Drop: Tractor/2 Trailers (An 8 psi drop within 2 minutes is permissible for either service reservoir) ☐   ☐

COMPLETE ALL NECESSARY REPAIRS BEFORE PROCEEDING TO TEST 3; SEE CHECK LIST 2 FOR COMMON CORRECTIONS.

Check List 2

EXCESSIVE LEAKAGE IN THE SUPPLY SIDE OF THE PNEUMATIC SYSTEM, COULD BE CAUSED BY:

NOTE: A leak detector or soap solution will aid in locating the faulty component

1. Supply lines and fittings (tighten)   ☐   ☐
2. Low pressure indicator(s)   ☐   ☐
3. Relay valves (antilock modulators)   ☐   ☐
4. Relay valve   ☐   ☐
5. Dual circuit foot brake valve   ☐   ☐
6. Trailer control valve   ☐   ☐
7. Park control valve   ☐   ☐
8. Tractor protection valve   ☐   ☐
9. Spring brake actuators   ☐   ☐
10. Safety pop-off valve in supply reservoir   ☐   ☐
11. Governor   ☐   ☐
12. Compressor discharge valves   ☐   ☐

*RETEST TO CHECK OUT ALL ITEMS REPAIRED OR REPLACED

Test 3
LEAKAGE SERVICE AIR DELIVERY

GOVERNOR CUT-OUT PRESSURE, ENGINE OFF
PARKING BRAKES RELEASED

1. Hold a service brake application   ☐   ☐
2. Allow pressure to stabilize for 1 minute; then begin timing for 2 minutes while observing the dash gauges
   - A. Pressure Drop: Single Vehicle (A 4 psi drop within 2 minutes is permissible for either service reservoir)   ☐   ☐
   - B. Pressure Drop: Tractor/Trailer (A 6 psi drop within 2 minutes is permissible for either service reservoir)   ☐   ☐
   - C. Pressure Drop: Tractor/2 Trailers (An 8 psi drop within 2 minutes is permissible for either service reservoir) ☐   ☐
3. Check brake chamber push rod travel (Refer to chart below for allowable tolerances)   ☐   ☐

Brake Chamber Size	Maximum Stroke Before Readjustment
12	1 1/4"
16	1 3/4"
20	1 3/4"
24	1 3/4"
30	2"

4. Check the angle formed between the brake chamber push rod and slack adjuster arm. (Should be 90° in the fully applied position)   ☐   ☐

COMPLETE ALL NECESSARY REPAIRS BEFORE PROCEEDING TO TEST 4; SEE CHECKLIST 3 FOR COMMON CORRECTIONS

Check List 3

Leakage in the service side of the pneumatic system, could be caused by:

NOTE: A leak detector or soap solution will aid in locating the defective component

	OK	NOT OK
1. Service lines and fittings (tighten)	☐	☐
2. Trailer control valve	☐	☐
3. Stoplight switch	☐	☐
4. Brake chamber diaphragms	☐	☐
5. Tractor protection valve	☐	☐
6. Relay valves (antilock modulators)	☐	☐
7. Service brake valve	☐	☐
8. Front axle ratio valve (optional)	☐	☐
9. Inverting relay spring brake control valve (optional) straight trucks and busses	☐	☐
10. Double check valve.	☐	☐

*RETEST TO CHECK OUT ALL ITEMS REPAIRED OR REPLACED

Test 4
MANUAL EMERGENCY SYSTEM

GOVERNOR CUT-OUT PRESSURE, ENGINE IDLING 600-900 RPM

FOR STRAIGHT TRUCKS, BUSES, AND BOBTAIL TRACTOR:

1. Manually operate the park control valve and note that parking brakes apply and release promptly. ☐ ☐

FOR TRACTOR/TRAILER COMBINATIONS:

1. Manually operate tractor protection control valve (trailer supply valve). The trailer brakes should apply and release promptly as control button is pulled out and pushed in. ☐ ☐
2. Manually operate system park control (usually yellow diamond button) and check that all parking brakes (tractor and trailer) apply promptly. ☐ ☐

COMPLETE ALL NECESSARY REPAIRS BEFORE PROCEEDING TO TEST 5; SEE CHECKLIST 4 FOR COMMON CORRECTIONS

Check List 4

If sluggish performance is noted in either test, check for:

	OK	NOT OK
1. Dented or kinked lines	☐	☐
2. Improperly installed hose fitting	☐	☐
3. A defective relay emergency valve	☐	☐
4. A defective modulator(s)	☐	☐

If the trailer brakes do not actuate and the trailer supply line remains charged, check the:

	OK	NOT OK
1. Tractor protection control	☐	☐
2. Trailer spring brake valve	☐	☐

*RETEST TO CHECK OUT ALL ITEMS REPAIRED OR REPLACED

Test 5

AUTOMATIC EMERGENCY SYSTEM

	OK	NOT OK
GOVERNOR CUT-OUT PRESSURE, ENGINE OFF		
1. Drain secondary circuit reservoir to 0 psi.		
A. Primary circuit reservoir should not lose pressure	☐	☐
B. On combination vehicles, the trailer air system should remain charged	☐	☐
C. Tractor and trailer brakes should not apply automatically	☐	☐
2. With no air pressure in the secondary circuit reservoir make a brake application.		
A. Rear axle brakes should apply and release	☐	☐
B. On combination vehicles the trailer brakes should also apply and release	☐	☐
C. The stop lamps should light	☐	☐
3. Slowly drain the primary circuit reservoir pressure.		
A. Spring brake push pull valve should pop out between 35 and 45 psi.	☐	☐
B. Tractor protection valve should close between 45 and 20 psi and trailer supply hose should be exhausted	☐	☐
C. Trailer brakes should apply after tractor protection closes	☐	☐
4. Close drain cocks, recharge system, and drain rear axle reservoir to 0 psi.		
A. Secondary circuit reservoir should not lose pressure	☐	☐
B. On combination vehicles the trailer air system should remain charged	☐	☐
5. With no air pressure in the primary circuit reservoir, make a brake application.		
A. Front axle brakes should apply and release	☐	☐
B. On combination vehicles the trailer brakes should also apply and release	☐	☐
C. If the vehicle is equipped with an inverting relay spring brake control valve, the rear axle brakes should also apply and release		

Check List 5

If the vehicle fails the previous tests outlined, then check the following:

	OK	NOT OK
1. Fittings	☐	☐
2. Kinked hose or tubing	☐	☐
3. Single check valves	☐	☐
4. Double check valves	☐	☐
5. Tractor protection valve	☐	☐
6. Tractor protection control valve	☐	☐
7. Parking control valve	☐	☐
8. Relay valves (antilock modulators)	☐	☐
9. Trailer spring brake control valve	☐	☐
10. Inverting relay spring brake control valve (optional) straight trucks and buses	☐	☐

*RETEST TO CHECK OUT ALL ITEMS REPAIRED OR REPLACED

STUDENT SELF-EVALUATION

Check	Level	Competency	Comments
	4	Mastered task	
	3	Competent but need further help	
	2	Needed a lot of help	
	0	Did not understand the task	

INSTRUCTOR EVALUATION

Check	Level	Competency	Comments
	4	Mastered task	
	3	Competent but needs further help	
	2	Requires more training	
	0	Unable to perform task	

JOB SHEET 31.3

Name _____ Station _____ Date _____

Use Colored Markers to Plumb in a Trailer Air Brake Circuit Using an OEM Schematic.

Performance Objective(s): Using colored markers, use an OEM schematic to plumb in the brake components on a tandem axle highway trailer.

ASE Education Foundation Correlation

This job sheet addresses the following ASE Education Foundation task(s):

III. Brakes
A. General
1. Research vehicle service information, including fluid type, vehicle service history, service precautions, and technical service bulletins. (IMMR, TST, MTST) P-1.
2. Identify brake system components and configurations (including air and hydraulic systems, parking brake, power assist, and vehicle dynamic brake systems). (IMMR, TST, MTST) P-1.

B. Air Brakes: Air Supply and Service Systems
3. Demonstrate knowledge and understanding of air supply and service system components and operations. (TST, MTST) P-1.

Tools and Materials: A standard highway trailer equipped with air brakes; plus a set of markers with at least three colors, preferably red, green, and blue

Protective Clothing: Standard shop apparel including coveralls or shop coat, safety glasses, and safety footwear

PROCEDURE
Ensure that the shop LOTO is observed.

Use Figure 31–7, which has the images of the hardware on a trailer air brake circuit, to connect the images using color codes as follows:

- Signal air (remember, the signal may be sourced from two circuits. Hint: use green or yellow)
- Trailer supply air (park/emergency circuit: use red for this)
- Hold-off air (use blue for this)
- Service application air (use green for this)

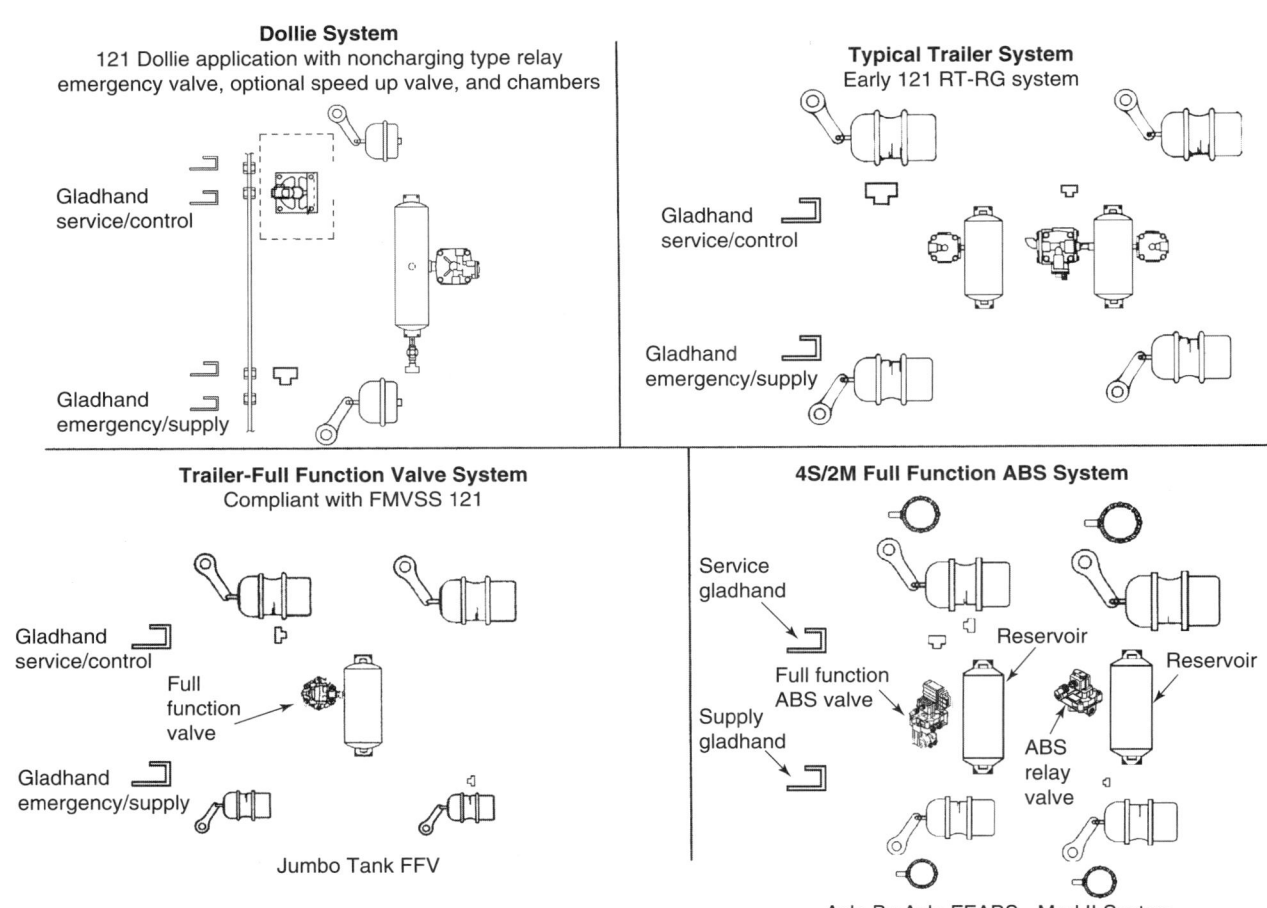

FIGURE 31-7 Use colored markers to plumb in the trailer brake circuit components. Suggestion: Use red for the park/emergency circuit, green for the service circuit, and blue for the hold-off circuit

STUDENT SELF-EVALUATION

Check	Level	Competency	Comments
	4	Mastered task	
	3	Competent but need further help	
	2	Needed a lot of help	
	0	Did not understand the task	

INSTRUCTOR EVALUATION

Check	Level	Competency	Comments
	4	Mastered task	
	3	Competent but needs further help	
	2	Requires more training	
	0	Unable to perform task	

JOB SHEET 31.4

Name _____ Station _____ Date _____

Use Colored Markers to Plumb in a Tractor Air Brake Circuit using an OEM Schematic.

Performance Objective(s): Using colored markers, use an OEM schematic or a parked tractor to plumb in the brake components on a tandem axle highway trailer.

ASE Education Foundation Correlation

This job sheet addresses the following ASE Education Foundation task(s):

III. Brakes
  A. General
  1. Research vehicle service information, including fluid type, vehicle service history, service precautions, and technical service bulletins. (IMMR, TST, MTST) P-1.
  2. Identify brake system components and configurations (including air and hydraulic systems, parking brake, power assist, and vehicle dynamic brake systems). (IMMR, TST, MTST) P-1.

  B. Air Brakes: Air Supply and Service Systems
  3. Demonstrate knowledge and understanding of air supply and service system components and operations. (TST, MTST) P-1.

Tools and Materials: A standard highway tractor equipped with air brakes; plus a set of markers with at least three colors, preferably red, green, and blue

Protective Clothing: Standard shop apparel including coveralls or shop coat, safety glasses, and safety footwear

PROCEDURE

Use Figure 31–8, which has the images of the hardware on a tractor air brake circuit, to connect the images using color codes as follows:

- Supply circuit air (use blue for this)
- Primary circuit (use green for this)
- Secondary circuit (use red for this)
- Trailer signal air (remember, this has two possible sources)
- Trailer supply air

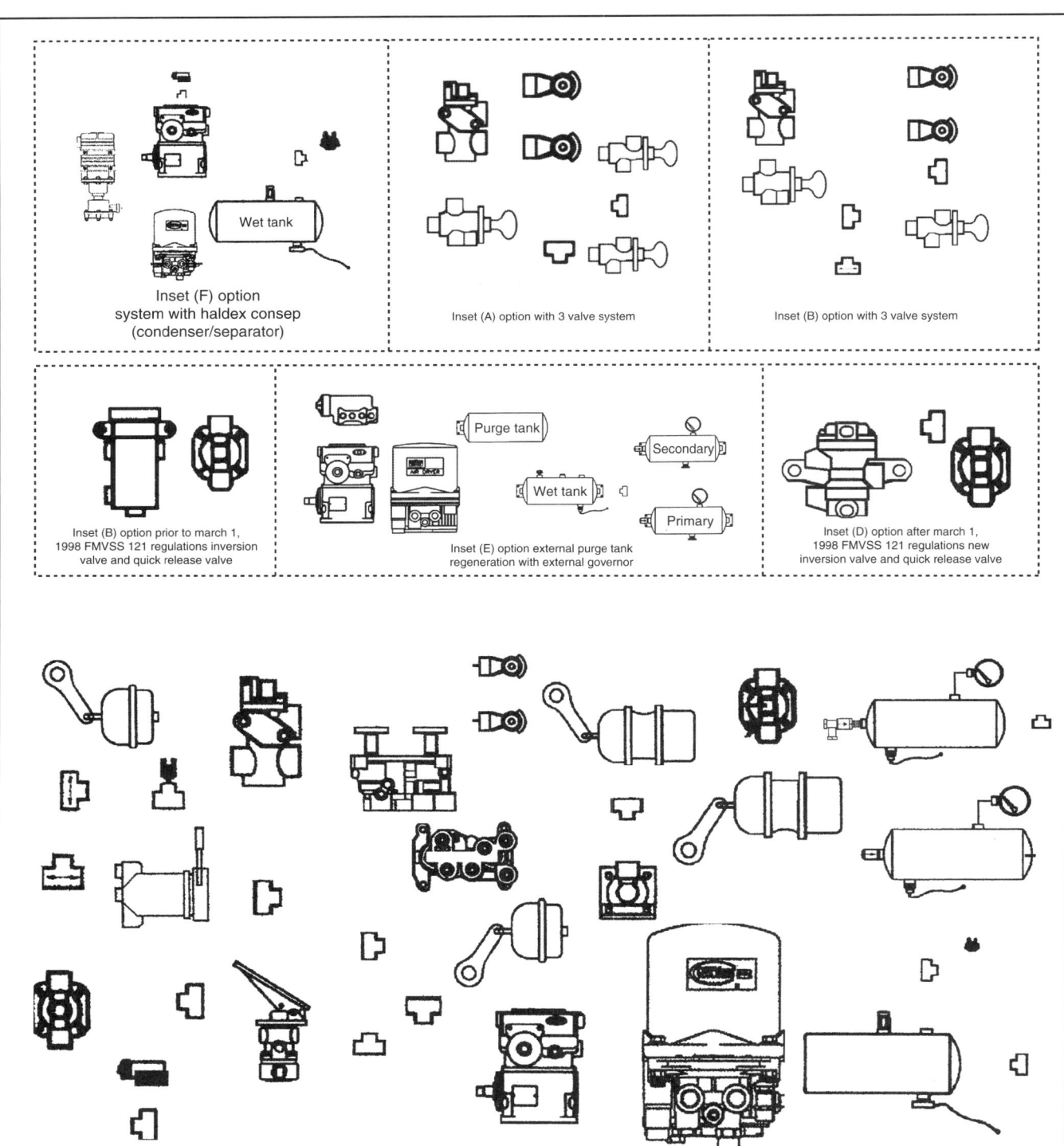

FIGURE 31-8 Tractor air brake circuit components: Use color markers to code and connect the components following the instructions in the worksheet

Task completed _____

STUDENT SELF-EVALUATION

Check	Level	Competency	Comments
	4	Mastered task	
	3	Competent but need further help	
	2	Needed a lot of help	
	0	Did not understand the task	

INSTRUCTOR EVALUATION

Check	Level	Competency	Comments
	4	Mastered task	
	3	Competent but needs further help	
	2	Requires more training	
	0	Unable to perform task	

ONLINE TASKS

Use your Bookmarks folder to identify the major truck brake system OEMs. At least one of them makes its complete brake troubleshooting service manual available online (free download). Identify this OEM and download the service manual. If you write this to a disk or thumb drive it can be an invaluable shop reference. Following are some pointers to get you started:

1. Bendix Brakes
2. Haldex Brakes
3. MGM Brake Actuators

STUDY TIPS

Identify 5 key points in Chapter 31. Try to be as brief as possible.

Key point 1 _____

Key point 2 _____

Key point 3 _____

Key point 4 _____

Key point 5 _____

32 Vehicle Chassis Frame

Objectives

After reading this chapter, you should be able to:
- Describe the construction of a chassis frame of a heavy-duty truck.
- Define the terms *yield strength, section modulus* (SM), and *resist bend moment* (RBM).
- List the materials from which frame rails are made and describe the characteristics of each.
- Explain the elements of frame construction.
- Describe the different ways that frame damage can occur as a result of impact and overloading.
- Perform some basic chassis frame alignment checks and project a frame-to-floor diagram.
- Describe the various categories of frame damage, including diamond, twist, sidesway, sag, and bow.
- Explain how the chassis frame, side rails, and cross members can be repaired without altering the frame dynamics.
- Specify the appropriate welding methods for repairing tempered and mild steel frame members.
- List some guidelines to follow when using frame repair hardware.

PRACTICE QUESTIONS

1. Heavy-duty Class 8 trucks use tempered steel-alloy frame rails with a yield strength rating of:
    a. 20,000 to 70,000 psi
    b. 30,000 to 80,000 psi
    c. 110,000 psi
    d. 150,000 psi

2. Which of the following best defines the term *section modulus* (SM)?
    a. It expresses shape and mounting geometry for purposes of calculating RBM.
    b. It expresses the material strength in an RBM calculation.
    c. It defines how easily a frame rail is deflected.
    d. It is the product of RBM and material yield strength.

3. What will be the effect of adding cross members to a heavy-duty truck ladder-type frame rail?
    a. increase in SM
    b. increase in RBM
    c. higher resistance to frame twist
    d. higher load-carrying ability

4. Which of the following methods will determine exactly what the frame material is for purposes of repair and modification?
    a. sales data book and VIN
    b. online data hub and VIN
    c. referencing the factory lineset ticket
    d. all of the above

5. Technician A says that when using an L reinforcement to a frame rail following a repair, the SM is changed. Technician B says that a fishplate L or inverted L reinforcement of a frame rail results in an increase in RBM. Who is correct?
   a. Technician A only
   b. Technician B only
   c. both A and B
   d. neither A nor B
6. Damage to a tractor chassis frame may be an indication of overloading the vehicle. Overloading can be caused by any of the following except:
   a. locating the fifth wheel at its rearmost position
   b. uneven location of the trailer load
   c. improper air pressure on a trailer lift axle
   d. exceeding the gross vehicle weight rating
7. All of the following are correct procedures to be observed when drilling holes in a heavy-duty tempered frame rail except:
   a. drilled bolt holes should be final-sized using a taper reamer
   b. drilling more than two holes on a vertical line
   c. drilling in the dead center of the web
   d. drilling in the drop portion of drop frame rails
8. Technician A says that when using U-bolts and clamps to fasten vehicle bodies to a tempered frame rail, steel pads should be welded to the side rail flanges. Technician B notches frame side rail flanges in order to obtain a perfect U-bolt fit. Who is correct?
   a. Technician A only
   b. Technician B only
   c. both Technicians A and B
   d. neither Technician A nor B
9. Which of the following are correct statements about RBM?
   a. It is the product of SM and material strength.
   b. The higher the RBM, the stronger the frame.
   c. RBM is the most meaningful measurement of frame rail strength.
   d. all of the above
10. When diagnosing a suspected frame misalignment condition on a tractor, which would be the correct sequence in which to perform the following checks?
    a. Check front-end alignment.
    b. Check bogie alignment.
    c. Check tire inflation.
    d. Project a frame-to-floor diagram.

JOB SHEET 32.1

Name _____ Station _____ Date _____

Project a Frame Diagram.

Performance Objective(s): Perform a frame-to-floor diagram to determine whether a frame has been twisted, diamonded, or otherwise damaged.

ASE Education Foundation Correlation

This job sheet addresses the following ASE Education Foundation task(s):

IV. Suspension and Steering Systems
   A. General
   1. Research vehicle service information, including fluid type, vehicle service history, service precautions, and technical service bulletins. (IMMR, TST, MTST) P-1.
   H. Frame and Coupling Devices
   2. Inspect frame and frame members for cracks, breaks, corrosion, distortion, elongated holes, looseness, and damage. (IMMR) P-1.
   2. Inspect frame and frame members for cracks, breaks, corrosion, distortion, elongated holes, looseness, and damage; determine needed action. (TST, MTST) P-1.
   3. Inspect frame hangers, brackets, and cross members. (IMMR) P-3.
   3. Inspect, install, and/or replace frame hangers, brackets, and cross members; determine needed action. (TST, MTST) P-3.

Tools and Materials: A highway tractor or straight truck; a 24-ft. (8-meter) tape measure; a plumb bob and chalk line; and a smooth, even, concrete shop floor

Protective Clothing: Standard shop apparel including coveralls or shop coat, safety glasses, and safety footwear

PROCEDURE
Ensure that the shop LOTO is observed.

Some OEMs provide instructions and recommended reference points on the chassis for this procedure, so attempt to locate this information before beginning. If it cannot be located, identify some reference points on the frame and suspension and use these. It is best to avoid frame fasteners as reference points.

1. Clean the shop floor thoroughly and then pull in the truck to be tested, running it backward and forward a couple of times to ensure that there is no lateral loading on either the frame or suspension.

   Task completed _____

2. Locate four sets of reference points on either side of the frame (eight in all). The first pair of reference points should be at the front of the truck and the other three more or less evenly spaced, with the final pair at the back of the truck. On each reference point, drop a plumb bob to the floor, settle it, and make a clearly visible reference mark on the floor with chalk.

   Task completed _____

3. When the eight reference points have been marked on the floor, carefully back the truck out of the shop. Now, using the chalk line, link the reference points longitudinally, laterally, and crisscross. The crisscross should be through each set of reference points (there will be four rectangular boxes on the floor) and, finally, crisscross from front left to right rear and front right to left rear. Take a look at the center crossing point of the crisscross chalk lines and determine whether the frame is properly trued. Most OEMs produce specifications on frame straightness, but, generally, any measurement exceeding ⅛ inch (3 mm) off true indicates a problem.

   Task completed _____

4. If a frame is measured to be out of specification, investigate what could have caused the condition and what could be done to repair it.

Task completed _____

STUDENT SELF-EVALUATION

Check	Level	Competency	Comments
	4	Mastered task	
	3	Competent but need further help	
	2	Needed a lot of help	
	0	Did not understand the task	

INSTRUCTOR EVALUATION

Check	Level	Competency	Comments
	4	Mastered task	
	3	Competent but needs further help	
	2	Requires more training	
	0	Unable to perform task	

ONLINE TASKS

Get online and check out the http://www.beeline-co.com site. Make a note of the type of equipment required to repair truck frames. Next identify the closest specialty truck frame repair shop to your town: There are not too many, so it may be some distance from you.

STUDY TIPS

Identify 5 key points in Chapter 32. Try to be as brief as possible.

Key point 1 _____

Key point 2 _____

Key point 3 _____

Key point 4 _____

Key point 5 _____

33 Heavy-Duty Truck Trailers

Objectives

After reading this chapter, you should be able to:

- Describe what is meant by semitrailers and full-trailers.
- Identify the various different tractor/trailer and train combinations.
- Describe what is meant by full frame, unibody, and monocoque.
- Explain the various types of hitching mechanisms used and their effects on tractor/trailer or train designation.
- Describe the design characteristics of the dry van, reefer, flatbed, tanker, and other types of highway trailer.
- Explain the operating principles of a reefer trailer.
- Identify the changes required for EPA tier IV (2013) compliant reefer engines.
- List the refrigerants currently used by reefer trailers.
- Outline some common trailer maintenance practices.
- Use a MUTT to diagnose trailer problems.

PRACTICE QUESTIONS

1. Technician A says that fiberglass-reinforced plywood (FRP) trailer walls are heavier than aluminum but tend to outlast aluminum. Technician B says that aluminum sidewalls are cheaper than FRP and, therefore, are more common. Who is correct?
   a. Technician A only
   b. Technician B only
   c. both A and B
   d. neither A nor B

2. Which of the following statements is true?
   a. Increasing the air pressure in a trailer lift axle decreases the percentage of weight carried by the trailer.
   b. Moving a fifth wheel forward on a tractor decreases trailer load.
   c. Moving the trailer bogies forward decreases the load carried by the trailer.
   d. none of the above

3. Technician A says that when the floor rots out in a unibody van trailer, the frame strength is reduced and the vehicle should not be driven. Technician B says that a damaged door post in a unibody van trailer reduces the frame strength. Who is correct?
   a. Technician A only
   b. Technician B only
   c. both A and B
   d. neither A nor B

4. What does the term *reefer* mean?
   a. an insulated freight van equipped with a refrigeration unit
   b. a drophead flatbed
   c. a curtained stake-bed trailer
   d. any trailer equipped with fruit doors

5. Which of the following trailer types would be most likely to be used to haul aggregates?
   a. tank trailer
   b. freight van
   c. flatbed trailer
   d. dump trailer

6. Technician A says that the difference between trailer-specific tires and tractor tires is that trailer tires have a shallower tread. Technician B says that the difference between trailer-specific tires and tractor tires is that trailer tires are lighter in weight. Who is correct?
   a. Technician A only
   b. Technician B only
   c. both A and B
   d. neither A nor B

7. Which of the following is true of trailer inspection maintenance?
   a. It can be performed less frequently than on tractors.
   b. Less maintenance is required because DOT inspections are less tough on trailers.
   c. It should be periodically scheduled according to linehaul miles.
   d. Trailers are less likely to be a hazard on the road than tractors, so they require fewer inspections.

8. Which of the following are benefits of trailer-specific tires?
   a. better retreadability
   b. better fuel mileage
   c. longer life, due to the three-groove design
   d. all of the above

9. What type of dump trailer is designed to handle the toughest hauling jobs?
   a. rock dumps
   b. slant front bathtub dumps
   c. bottom hopper dump
   d. frameless bathtub dumps

10. Technician A says that most current reefer units are designed for R-134a use. Technician B says that unlike automobile refrigerants, trailer refrigerants are more environmentally sound and can be exhausted to atmosphere. Who is correct?
    a. Technician A only
    b. Technician B only
    c. both Technicians A and B
    d. neither Technician A nor B

JOB SHEET 33.1

Name _____ Station _____ Date _____

Check Alignment on a Van Trailer.

Performance Objective(s): Check the bogie-to-van body alignment on a trailer to check for a dog-tracking condition.

ASE Education Foundation Correlation
This job sheet addresses the following ASE Education Foundation task(s):

IV. Suspension and Steering Systems
   A. General
   1. Research vehicle service information, including fluid type, vehicle service history, service precautions, and technical service bulletins. (IMMR, TST, MTST) P-1.

   E. Suspension Systems
   2. Inspect leaf springs, center bolts, clips, pins, bushings, shackles, U-bolts, insulators, brackets, and mounts; determine needed action. (IMMR) P-1.
   2. Inspect, repair, and/or replace leaf springs, center bolts, clips, pins, bushings, shackles, U-bolts, insulators, brackets, and mounts; determine needed action. (TST, MTST) P-1.

   H. Frame and Coupling Devices
   2. Inspect frame and frame members for cracks, breaks, corrosion, distortion, elongated holes, looseness, and damage. (IMMR) P-1.
   2. Inspect frame and frame members for cracks, breaks, corrosion, distortion, elongated holes, looseness, and damage; determine needed action. (TST, MTST) P-1.

Tools and Materials: A unibody tandem axle van-type cargo trailer, alignment equipment (including laser if available), a bazooka, a plumb bob, a chalk line, a tape measure, a clean concrete shop floor, and standard shop hand and power tools

Protective Clothing: Standard shop apparel including coveralls or shop coat, safety glasses, and safety footwear

PROCEDURE
Ensure that the shop LOTO is observed.

There are two acceptable methods of performing this procedure. The method outlined here is low tech, but similar results can be obtained in about a quarter of the time by using laser alignment instruments.

1. Make sure that the trailer van to be tested is unloaded. Clean the shop floor thoroughly and pull the trailer in. Run it back and forth a couple of times to relieve any lateral thrust load on the suspension before parking. Block the trailer wheels, uncouple the tractor, and then set the trailer height using dolly legs to the specified bolster plate height.

   Task completed _____

2. Locate three index points on either side of the van body; these should be the front, middle, and back of the trailer.

   Task completed _____

3. Use the plumb bob to project the index points to the floor and mark.

   Task completed _____

4. Locate two index points on either side of the suspension bogie, again front and back, and the same on either side. Use the plumb bob to project these to the floor.

   Task completed _____

5. Next fit axle extenders to each trailer axle end.

   Task completed _____

6. Mount the bazooka to the kingpin on the upper coupler and ensure that the instrument level indicates that it is set true. Measure the bazooka-to-axle extender dimensions and record them. Measure the axle extender-to-extender dimension on each side of the trailer and record it.

   Task completed _____

7. Remove the bazooka and axle extenders. Couple the tractor to the trailer and remove them from the shop.

   Task completed _____

8. Now use the chalk line to make a comparison between the bogie rectangle and the van body. Note the location of the crisscross centerpoints.

   Task completed _____

9. Check the geometry created on the floor by projecting the bogie and van body diagrams with the measurements made with the bazooka and axle extenders.

   Task completed _____

10. Check the trailer OEM alignment specifications and determine whether your measurements meet them.

    Task completed _____

STUDENT SELF-EVALUATION

Check	Level	Competency	Comments
	4	Mastered task	
	3	Competent but need further help	
	2	Needed a lot of help	
	0	Did not understand the task	

INSTRUCTOR EVALUATION

Check	Level	Competency	Comments
	4	Mastered task	
	3	Competent but needs further help	
	2	Requires more training	
	0	Unable to perform task	

JOB SHEET 33.2

Name _____ Station _____ Date _____

Inspect Trailer Landing Gear.

Performance Objective(s): Check the operation of trailer landing gear and note any problems.

ASE Education Foundation Correlation

This job sheet addresses the following ASE Education Foundation task(s):

IV. Suspension and Steering Systems
  A. General
  1. Research vehicle service information, including fluid type, vehicle service history, service precautions, and technical service bulletins. (IMMR, TST, MTST) P-1.

  H. Frame and Coupling Devices
  2. Inspect frame and frame members for cracks, breaks, corrosion, distortion, elongated holes, looseness, and damage. (IMMR) P-1.
  2. Inspect frame and frame members for cracks, breaks, corrosion, distortion, elongated holes, looseness, and damage; determine needed action. (TST, MTST) P-1.
  3. Inspect frame hangers, brackets, and cross members. (IMMR) P-3.
  3. Inspect and install frame hangers, brackets, and cross members; determine needed action. (TST) P-3.
  3. Inspect, install, and/or replace frame hangers, brackets, and cross members; determine needed action. (MTST) P-3.

Tools and Materials: A unibody tandem axle trailer, with a mechanical landing gear and a two-speed gearbox. Standard shop equipment and lubricants.

Protective Clothing: Standard shop apparel including coveralls or shop coat, safety glasses, and safety footwear

PROCEDURE
Ensure that the shop LOTO is observed.

This procedure should only be performed on landing gear that not categorized as 5-year or 10-year No-Lube. Make sure that the trailer is either unloaded or lightly loaded. Clean the shop floor thoroughly and pull the trailer in.

1. Park the rig, apply the parking brakes, and leaving the trailer coupled to the tractor, block the trailer wheels.

   Task completed _____

2. Check for ovalling of the sidewall axle through holes.

   Task completed _____

3. Ensure axle roll pins are in place and secure.

   Task completed _____

4. Check for vertical movement of the inner leg. If present, this may indicate a damaged or worn thrust bearing.

   Task completed _____

5. Check the shifter shaft for smooth inboard/outboard movement and ensure that it fully engages in high and low gear ratios.

   Task completed _____

6. Clean the shifter shaft with clean emery cloth and apply a light oil lubricant.

   Task completed _____

7. Check the gearbox bushings to ensure that they are not damaged or ovalled.

    Task completed _____

8. Check the gearbox crossshaft to ensure that it moves freely.

    Task completed _____

9. Fully retract the inner leg. Extend the leg using the high gear ratio for two full cranks: then apply approximately ½ pound (0.25 kg.) of chassis grease into the lower zerk fitting in the outer tube. Retract the inner leg.

    Task completed _____

10. Cycle the landing gear through a full retraction and extension cycle to circulate the grease.

    Task completed _____

11. Apply approximately ½ pound (0.25 kg.) of chassis grease into the top zerk fitting in the outer tube.

    Task completed _____

12. Apply approximately ½ pound (0.25 kg.) of chassis grease into the gearbox while cycling it through one full extension and retraction cycle. Ensure that the drain hole in the bottom of the gearbox is clear.

    Task completed _____

13. For a C-inspection, disassemble the gearbox, clean out the old grease and repack with OEM-specified weight of grease. Remove the top cover on each leg, clean out old grease and repack with OEM-specified weight of grease.

    Task completed _____

STUDENT SELF-EVALUATION

Check	Level	Competency	Comments
	4	Mastered task	
	3	Competent but need further help	
	2	Needed a lot of help	
	0	Did not understand the task	

INSTRUCTOR EVALUATION

Check	Level	Competency	Comments
	4	Mastered task	
	3	Competent but needs further help	
	2	Requires more training	
	0	Unable to perform task	

ONLINE TASKS

Use a search engine and identify five semi-trailer manufacturers and log them into your Bookmark folders. To get you started, search the Internet for Trailmobile, Tremcar, and Heil Trailers. Make a note of any specialty areas of the industry that each manufacturer services. Also check out the Truck Trailer Manufacturer's Association (TTMA).

1. _____
2. _____
3. _____
4. _____
5. _____

STUDY TIPS

Identify 5 key points in Chapter 33. Try to be as brief as possible.

Key point 1 _____

Key point 2 _____

Key point 3 _____

Key point 4 _____

Key point 5 _____

34 Fifth Wheels and Coupling Systems

Objectives

After reading this chapter, you should be able to:

- Describe the different types of fifth wheels used on tractors.
- Outline the operating principles of the Holland, Fontaine, and JOST fifth wheels.
- Understand the importance of correctly locating the fifth wheel on the tractor.
- Describe the locking principles of each type of fifth wheel.
- Outline the procedure required to couple and uncouple a fifth wheel.
- Inspect and service the common types of fifth wheels.
- Describe the procedure required to overhaul an SAF-Holland fifth wheel.
- Identify the overhaul procedure required of some common fifth wheels.
- Outline the importance of properly positioning sliding fifth wheels and explain how this alters chassis weight-over-axle.
- Define high hitch and outline what is required to avoid it.
- Outline the function of the kingpin and upper coupler assembly.
- Describe the operating principles of pintle hooks, ball hitches, and draw bars.
- Inspect and service a pintle hook assembly.
- Identify OOS factors of HD coupling devices.

PRACTICE QUESTIONS

1. Technician A says that minor cracks in a cast-steel fifth wheel deck can be repair-welded using low-hydrogen electrodes. Technician B says that worn trailer kingpins can be built up with weld in position on the upper coupler assembly and returned to service. Who is correct?
   a. Technician A only
   b. Technician B only
   c. both A and B
   d. neither A nor B

2. Which type of fifth wheel would be recommended for use on a tanker trailer, double B-train?
   a. stabilized
   b. fully oscillating
   c. semi-oscillating
   d. compensating

3. Which type of fifth wheel is capable of altering the amount of weight over a tractor bogie assembly?
   a. semi-oscillating
   b. oscillating
   c. sliding
   d. rigid

4. Which of the following is a factor in determining the fifth wheel height?
   a. trailer height
   b. tractor frame height
   c. tire clearance
   d. all of the above

5. Which of the following is true of the angle-on-frame mount as compared to the plate mount?
   a. higher cost
   b. high torsional rigidity
   c. less torsional rigidity
   d. increased trailer torque effect

6. The correct sliding fifth wheel location should be based on:
   a. behind-cab clearance
   b. landing gear clearance
   c. weight distribution and load over the tractor bogies
   d. all of the above

7. Technician A says that less fore-and-aft articulation is possible as fifth wheel height decreases. Technician B says that off-highway use will normally require a higher fifth wheel capacity rating. Who is correct?
   a. Technician A only
   b. Technician B only
   c. both A and B
   d. neither A nor B

8. Which of the following lubricants should be used on a fifth wheel deck that requires lubricating?
   a. water-resistant, lithium-based grease
   b. light grease
   c. engine oil
   d. heavy-duty gear oil

9. Technician A replaces a Holland jaw pair if wear exceeds ⅛ inch (3 mm) at the kingpin bearing surface. Technician B replaces the buffing rubber, jaws, and lock if the shims total 3⁄16 inch (4.5 mm). Who is correct?
   a. Technician A only
   b. Technician B only
   c. both A and B
   d. neither A nor B

10. When the locking plungers of an air release sliding fifth wheel will not release to fore-and-aft movement of the assembly, Technician A checks the plungers for wear. Technician B checks the air cylinder for proper operation. Who is correct?
    a. Technician A only
    b. Technician B only
    c. both A and B
    d. neither A nor B

JOB SHEET 34.1

Name _____ Station _____ Date _____

Check and Adjust Play in a Holland Fifth Wheel.

Performance Objective(s): Use a kingpin test tool to check play in a Holland fifth wheel and adjust if necessary.

ASE Education Foundation Correlation
This job sheet addresses the following ASE Education Foundation task(s):

IV. Suspension and Steering Systems
  H. Frame and Coupling Devices
   1. Inspect, service, and/or adjust fifth wheel, pivot pins, bushings, locking mechanisms, mounting hardware, air lines, and fittings. (IMMR, TST, MTST) P-1.
   3. Inspect frame hangers, brackets, and cross members. (IMMR) P-3.
   3. Inspect and install frame hangers, brackets, and cross members; determine needed action. (TST) P-3.
   3. Inspect, install, and/or replace frame hangers, brackets, and cross members; determine needed action. (MTST) P-3.

Tools and Materials: A truck equipped with a Holland A or B fifth wheel, a kingpin test tool, and standard shop hand and power tools

Protective Clothing: Standard shop apparel including coveralls or shop coat, safety glasses, and safety footwear

PROCEDURE
Ensure that the shop LOTO is observed.

1. Insert a kingpin test dummy (such as the one shown in Figure 34–1) into the throat of the kingpin and ensure that the fifth wheel primary and secondary locks engage.

   Task completed _____

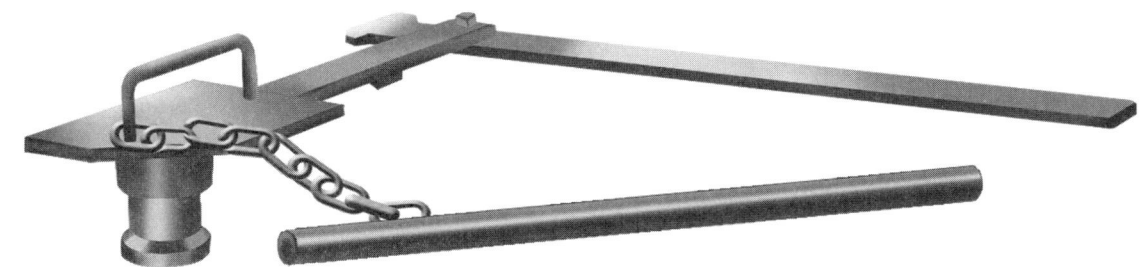

FIGURE 34–1 Kingpin lock tester

2. Check the amount of fore-and-aft play; up to ¼ inch (6 mm) play is permissible on a Holland fifth wheel. If more than ¼ inch play is evident, the fifth wheel should be adjusted.

   Task completed _____

3. On the Holland fifth wheel, the adjustment nut and washer are located at the front of the top plate; they slide in and out as the fifth wheel jaws engage. With the fifth wheel locked, the rubber bushing should fit snug enough to the plate so it can be just about rotated by hand. If the fifth wheel adjustment is loose, the bushing will be too tight to turn by hand.

   Task completed _____

4. To adjust the fifth wheel, turn the adjusting nut counterclockwise (CCW) to adjust for wear. Turn the nut CCW until you feel minimal fore-and-aft clearance. The adjusting nut is self-locking so no jam nut is required. After making an adjustment, lock and unlock the fifth wheel a couple of times to recheck the adjustment. Ensure that the bushing can be rotated. If this is backed out so that the bushing becomes loose, turn the adjusting nut clockwise (CW).

Task completed _____

5. Now test the adjustment by coupling again to the trailer. Ensure that the fifth wheel properly locks and releases.

Task completed _____

STUDENT SELF-EVALUATION

Check	Level	Competency	Comments
	4	Mastered task	
	3	Competent but need further help	
	2	Needed a lot of help	
	0	Did not understand the task	

INSTRUCTOR EVALUATION

Check	Level	Competency	Comments
	4	Mastered task	
	3	Competent but needs further help	
	2	Requires more training	
	0	Unable to perform task	

JOB SHEET 34.2

Name _____ Station _____ Date _____

Reposition a Sliding Fifth Wheel.

Performance Objective(s): Become familiar with the procedure used to reposition a sliding fifth wheel on a tractor unit.

ASE Education Foundation Correlation

This job sheet addresses the following ASE Education Foundation task(s):

IV. Suspension and Steering Systems
H. Frame and Coupling Devices
1. Inspect, service, and/or adjust fifth wheel, pivot pins, bushings, locking mechanisms, mounting hardware, air lines, and fittings. (IMMR, TST, MTST) P-1.
2. Inspect frame and frame members for cracks, breaks, corrosion, distortion, elongated holes, looseness, and damage. (IMMR) P-1.
2. Inspect frame and frame members for cracks, breaks, corrosion, distortion, elongated holes, looseness, and damage; determine needed action. (TST, MTST) P-1.
3. Inspect frame hangers, brackets, and cross members. (IMMR) P-3.
3. Inspect and install frame hangers, brackets, and cross members; determine needed action. (TST) P-3.
3. Inspect, install, and/or replace frame hangers, brackets, and cross members; determine needed action. (MTST) P-3.

Tools and Materials: A tractor equipped with either a manual or pneumatic release, sliding fifth wheel, and a trailer

Protective Clothing: Standard shop apparel including coveralls or shop coat, safety glasses, and safety footwear

PROCEDURE
Ensure that the shop LOTO is observed.

1. Park a coupled tractor/trailer combination in a straight line on level ground.

   Task completed _____

2. Place the trailer brakes in park.

   Task completed _____

3. If the tractor is equipped with an air slide release, switch the cab control valve into the unlocked position.

   Task completed _____

4. If the tractor is equipped with a manual release (reference Figure 34–2), pull the release lever until the locking plungers/wedges are completely disengaged from the sliding deck teeth.

   Task completed _____

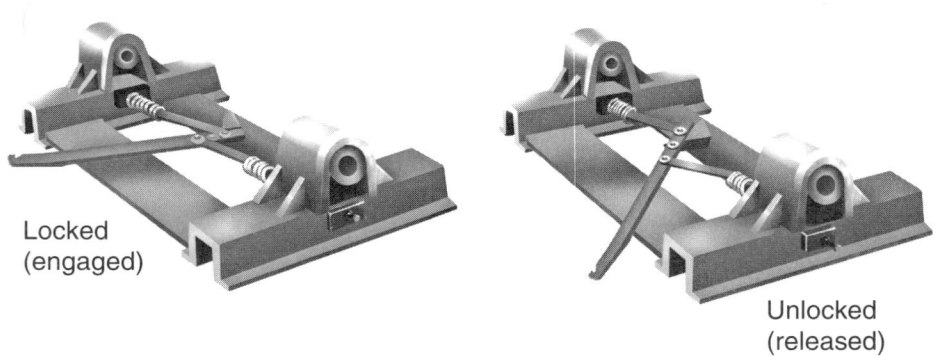

FIGURE 34–2 Manual slide release

5. Take a look at Figure 34–3 and make sure that the locking plungers/wedges are completely disengaged before proceeding.

Task completed _____

FIGURE 34–3 Locking plungers in locked and unlocked positions

6. If the plungers do not completely disengage, lowering the trailer landing gear may help relieve pressure on the plungers.

Task completed _____

7. Now release the tractor brakes, and gently nudge forward or backward to reposition the fifth wheel at the desired setting.

Task completed _____

8. When the new fifth wheel location has been determined, if an air slide release is used, switch the control valve back into the locked position to retract the plungers.

Task completed _____

9. If a manual slide release is used, trip the release arm to allow the plungers to retract.

Task completed _____

10. Visually check to see that both plungers are retracted and fully engaged to the teeth in the slide rails.

Task completed _____

11. If necessary, move the tractor a small amount to fully reengage the plungers back into the slide rail teeth.

Task completed _____

12. Raise the trailer landing gear to the fully retracted position.

Task completed _____

STUDENT SELF-EVALUATION

Check	Level	Competency	Comments
	4	Mastered task	
	3	Competent but need further help	
	2	Needed a lot of help	
	0	Did not understand the task	

INSTRUCTOR EVALUATION

Check	Level	Competency	Comments
	4	Mastered task	
	3	Competent but needs further help	
	2	Requires more training	
	0	Unable to perform task	

JOB SHEET 34.3

Name _____ Station _____ Date _____

Check and Adjust Play in a JOST Fifth Wheel.

Performance Objective(s): Use a kingpin test tool to check play in a JOST fifth wheel and adjust if necessary.

ASE Education Foundation Correlation
This job sheet addresses the following ASE Education Foundation task(s):

IV. Suspension and Steering Systems
  H. Frame and Coupling Devices
  1. Inspect, service, and/or adjust fifth wheel, pivot pins, bushings, locking mechanisms, mounting hardware, air lines, and fittings. (IMMR, TST, MTST) P-1.
  2. Inspect frame and frame members for cracks, breaks, corrosion, distortion, elongated holes, looseness, and damage. (IMMR) P-1.
  2. Inspect frame and frame members for cracks, breaks, corrosion, distortion, elongated holes, looseness, and damage; determine needed action. (TST, MTST) P-1.
  3. Inspect frame hangers, brackets, and cross members. (IMMR) P-3.
  3. Inspect and install frame hangers, brackets, and cross members; determine needed action. (TST) P-3.
  3. Inspect, install, and/or replace frame hangers, brackets, and cross members; determine needed action. (MTST) P-3.

Tools and Materials: A truck equipped with a standard JOST fifth wheel, a kingpin test tool, and standard shop hand and power tools

Protective Clothing: Standard shop apparel including coveralls or shop coat, safety glasses, and safety footwear

PROCEDURE
Ensure that the shop LOTO is observed.

1. Insert a kingpin test dummy (such as the one shown in Figure 34–1) into the throat of the kingpin and ensure that the fifth wheel primary and secondary locks engage.

   Task completed _____

2. Check the amount of fore-and-aft play. There should be no fore-and-aft play on a JOST fifth wheel. If fore-and-aft play is evident, the fifth wheel should be adjusted.

   Task completed _____

3. On the JOST fifth wheel, the adjustment mechanism is a bolt and locknut located on the right side, slightly to the rear of the centerline. To adjust, loosen the jam nut with a wrench.

   Task completed _____

4. If the adjustment is too loose, turn the adjustment screw CCW one full turn. Then retorque the jam nut. If the adjustment is too tight, the adjustment screw will have to be turned CW one full turn. Lock and unlock the unit several times to ensure that the adjustment is properly set.

   Task completed _____

5. Now test the adjustment by coupling again to the trailer. Ensure that the fifth wheel properly locks and releases.

   Task completed _____

STUDENT SELF-EVALUATION

Check	Level	Competency	Comments
	4	Mastered task	
	3	Competent but need further help	
	2	Needed a lot of help	
	0	Did not understand the task	

INSTRUCTOR EVALUATION

Check	Level	Competency	Comments
	4	Mastered task	
	3	Competent but needs further help	
	2	Requires more training	
	0	Unable to perform task	

ONLINE TASKS

Use a search engine and get each of the OEMs listed next logged into your bookmark/favorites folder. What can you discover about each OEM product from their online presentation?

1. SAF Holland
2. JOST fifth wheels
3. Fontaine

STUDY TIPS

Identify 5 key points in Chapter 34. Try to be as brief as possible.

Key point 1 _____

Key point 2 _____

Key point 3 _____

Key point 4 _____

Key point 5 _____

35 Heavy-Duty Heating, Ventilation, and Air-Conditioning Systems

Objectives

After reading this chapter, you should be able to:

- Understand the basic theory of heavy-duty truck air-conditioning systems.
- Outline the requirements of the Clean Air Act that apply to a heavy-duty truck air-conditioning system.
- List the five major components of a heavy-duty air-conditioning system and describe how each works in the operation of the system.
- Explain how the thermostatic expansion valve or orifice tube controls the flow of refrigerant to the evaporator.
- Identify the refrigerants used in heavy-duty truck air-conditioning systems.
- Identify the GWP characteristics of R-134a, R-1234yf, and CO_2.
- Describe the function of the main components in a typical heavy-duty air-conditioning system.
- Recognize the environmental and personal safety precautions that must be observed when working on air-conditioning systems.
- Identify air-conditioning testing and service equipment.
- Test an air-conditioning system for refrigerant leaks.
- Outline the procedure required to service a heavy-duty air-conditioning system.
- Describe the differences between R-134a and R-1234yf A/C system servicing procedures and equipment.
- Perform some simple diagnosis of air-conditioning system malfunctions.
- Outline the advantages of connecting air-conditioning management electronics to the chassis data bus and explain how to access the system.
- Explain how PWM-actuated controls are used to manage refrigerant circuit switches and move air doors in electronically managed HVAC systems.
- Describe how virtual switches are used in current HVAC systems.
- Identify the key J1939 source addresses (SAs) used to manage HVAC.
- Explain how a truck cab ventilation system operates.
- Describe the role a liquid-cooled heating system plays in a truck cab heating system.
- Describe some types of auxiliary heating and power units.

PRACTICE QUESTIONS

1. Technician A says that the latent heat of vaporization can be measured in British thermal units (Btus). Technician B says that the pressure acting on a liquid changes its boiling point. Who is correct?
   a. Technician A only
   b. Technician B only
   c. both A and B
   d. neither A nor B

2. Technician A says that a rotary vane compressor uses no sealing rings. Technician B says that a rotary compressor has an oil sump on its discharge side. Who is correct?
   a. Technician A only
   b. Technician B only
   c. both A and B
   d. neither A nor B

3. What type of electromagnetic clutch is most commonly used in truck air-conditioning (A/C) applications?
   a. rotating coil type
   b. stationary coil type
   c. split coil type
   d. centrifugal

4. The receiver/dryer performs which of the following functions?
   a. filters particles from the liquid refrigerant
   b. removes moisture from liquid refrigerant
   c. both a and b
   d. neither a nor b

5. Which of the following observations through an R-12 sight glass can indicate that the system is empty?
   a. oil streaking
   b. bubbles
   c. foam
   d. all of the above

6. Which of the following should be done before testing the operation of the A/C system?
   a. Check for deteriorated or cracked hoses.
   b. Check the refrigerant compressor's drive belt tension.
   c. Visually inspect for evidence of oil streaking around connections.
   d. all of the above

7. Which fuel is used to fire most APUs found on today's trucks?
   a. diesel fuel
   b. gasoline
   c. propane
   d. methyl hydrate

8. Which heat output rating APU should be used in a Class 8 truck equipped with a 60-inch (1.5 meter) bunk and operated in North American winter conditions?
   a. 25,000 Btus (26,376 kilojoules)
   b. 30,000 Btus (31,652 kilojoule)
   c. 40,000 Btus (42,202 kilojoule)
   d. 100,000 Btus (105,506 kilojoule)

9. Which type of blower fan is used in most truck heating and ventilation systems?
   a. spiral-rotor
   b. four-blade
   c. squirrel-cage
   d. five-blade

10. What is the medium used inside a cab heating system heat exchanger core?
    a. engine oil
    b. engine coolant
    c. refrigerant
    d. ram air

Test Exercises

1. Make a list of auxiliary heating units that could be found on a modern truck.

2. Make a list of currently used refrigerants in truck A/C and reefer units. See if you can determine which still have chlorine in them.

3. Research MY 2020 proposed refrigerants such as R-1234yf and R-422d and determine the ODP and GWP of each.

4. Pose arguments for and against adopting propane and CO_2 as mobile refrigerants in commercial trucks.

JOB SHEET 35.1

Name _____ Station _____ Date _____

Identify the Type of Air-Conditioning (A/C) System in a Vehicle.

Performance Objective(s): Determine the type of A/C system used in a vehicle by observation and using the OEM support literature.

ASE Education Foundation Correlation

This job sheet addresses the following ASE Education Foundation task(s):

VI. Heating, Ventilation and Air-Conditioning (HVAC)
A. General
1. Research vehicle service information, including refrigerant/oil type, vehicle service history, service precautions, and technical service bulletins. (IMMR, TST, MTST) P-1.
2. Identify heating, ventilation, and air-conditioning (HVAC) components and configuration. (IMMR, TST, MTST) P-1.
3. Use appropriate electronic service tool(s) and procedures to check, record, and clear diagnostic codes; interpret digital multimeter (DMM) readings. (IMMR) P-1.
3. Use appropriate electronic service tool(s) and procedures to diagnose problems; check, record, and clear diagnostic codes; interpret digital multimeter (DMM) readings. (TST, MTST) P-1.

C. Heating, Ventilation, and Engine Cooling Systems
2. Inspect HVAC system-heater ducts, doors, hoses, cabin filters, and outlets; determine needed action. (IMMR, TST, MTST) P-1.

Tools and Materials: A truck equipped with an A/C system and OEM support literature. For systems using R-1234yf which are currently being introduced, please consult OEM service literature

Protective Clothing: Standard shop apparel including coveralls or shop coat, safety glasses, and safety footwear

PROCEDURE
Ensure that the shop LOTO is observed.

1. Observe the dash controls. Describe the controls and make a note of how the system is managed, that is, by "smart" controls (electronic climate control) or by manual settings.

   Task completed _____

2. Open the hood. Make a note of any data plates or labels that relate to the A/C system. Identify the A/C system manufacturer.

   Task completed _____

3. Check the types of fittings used in the A/C system and identify the type of refrigerant used.

   Task completed _____

4. Identify what type of refrigerant oil is used in the system.

   Task completed _____

5. Check the operation of the system. Use a thermometer in the outlet duct closest to the evaporator and determine whether the system is performing to specifications.

   Task completed _____

STUDENT SELF-EVALUATION

Check	Level	Competency	Comments
	4	Mastered task	
	3	Competent but need further help	
	2	Needed a lot of help	
	0	Did not understand the task	

INSTRUCTOR EVALUATION

Check	Level	Competency	Comments
	4	Mastered task	
	3	Competent but needs further help	
	2	Requires more training	
	0	Unable to perform task	

JOB SHEET 35.2

Name _____ Station _____ Date _____

Identify the Chassis Location of A/C Components.

Performance Objective(s): Become familiar with the location of the major A/C components on a truck chassis.

ASE Education Foundation Correlation
This job sheet addresses the following ASE Education Foundation task(s):

VI. Heating, Ventilation, and Air Conditioning (HVAC)
A. General
1. Research vehicle service information, including refrigerant/oil type, vehicle service history, service precautions, and technical service bulletins. (IMMR, TST, MTST) P-1.
2. Identify heating, ventilation, and air-conditioning (HVAC) components and configuration. (IMMR, TST, MTST) P-1.
3. Use appropriate electronic service tool(s) and procedures to check, record, and clear diagnostic codes; interpret digital multimeter (DMM) readings. (IMMR) P-1.
3. Use appropriate electronic service tool(s) and procedures to diagnose problems; check, record, and clear diagnostic codes; interpret digital multimeter (DMM) readings. (TST, MTST) P-1.

B. Refrigeration System Components
1. Inspect A/C compressor drive belts, pulleys, and tensioners; verify proper belt alignment. (IMMR) P-1.
1. Inspect, remove, and replace A/C compressor drive belts, pulleys, and tensioners; verify proper belt alignment. (TST, MTST) P-1.
2. Check A/C system operation including system pressures; visually inspect A/C components for signs of leaks; check A/C monitoring system (if applicable). (IMMR. TST, MTST) P-1.
3. Inspect A/C condenser for airflow restrictions; determine needed action. (IMMR, TST, MTST) P-1.

C. Heating, Ventilation, and Engine Cooling Systems
1. Inspect engine cooling system and heater system hoses and pipes; determine needed action. (IMMR, TST, MTST) P-1.
2. Inspect HVAC system-heater ducts, doors, hoses, cabin filters, and outlets; determine needed action. (IMMR) P-1.
2. Inspect HVAC system heater ducts, doors, hoses, cabin filters, and outlets; determine needed action. (TST, MTST) P-1.
3. Identify the source of A/C system odors. (IMMR) P-2.
3. Identify the source of A/C system odors; determine needed act. (TST, MTST) P-2.

D. Operating Systems and Related Controls
1. Verify blower motor operation; confirm proper air distribution; confirm proper temperature control; determine needed action. (IMMR) P-1.
1. Verify HVAC system blower motor operation; confirm proper air distribution; confirm proper temperature control; determine needed action. (TST, MTST) P-1.

Tools and Materials: A truck equipped with an A/C system

Protective Clothing: Standard shop apparel including coveralls or shop coat, safety glasses, and safety footwear

PROCEDURE
Ensure that the shop LOTO is observed.

Note the location and the manufacturer name, if possible, for each component in the following table.
The components listed here may not appear on all A/C systems.

Component	Description of location, type, manufacturer, etc.
Compressor	
Evaporator	
Condenser	
Receiver/dryer	
TXV/orifice tube	
High-pressure line	
Low-pressure line	
Sight glass	
Accumulator	
Blower motor/fan	
APU	
HVAC ECU	
Thermostat	
Trinary valve	

Task completed _____

STUDENT SELF-EVALUATION

Check	Level	Competency	Comments
	4	Mastered task	
	3	Competent but need further help	
	2	Needed a lot of help	
	0	Did not understand the task	

INSTRUCTOR EVALUATION

Check	Level	Competency	Comments
	4	Mastered task	
	3	Competent but needs further help	
	2	Requires more training	
	0	Unable to perform task	

ONLINE TASKS

Use a search engine and check out what the EPA has to say about vehicle and specifically truck air-conditioning and trailer reefer systems. Make a note of three emerging trends in vehicle A/C systems.

1. _____
2. _____
3. _____

STUDY TIPS

Identify 5 key points in Chapter 35. Try to be as brief as possible.

Key point 1 _____

Key point 2 _____

Key point 3 _____

Key point 4 _____

Key point 5 _____